Highland Wildlife

Books by the same Author

At the Turn of the Tide

A Naturalist on Lindisfarne

Watching Sea-Birds

The Watcher and the Red Deer

Wildlife in Britain and Ireland

Highland Wildlife

Richard Perry

CROOM HELM LONDON

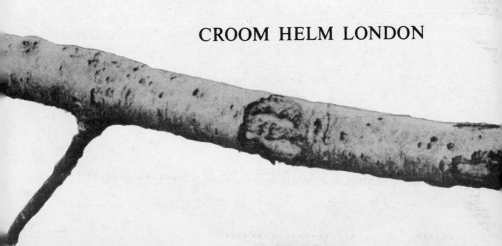

© 1979 Richard Perry
Croom Helm Ltd, 2–10 St John's Road, London SW11
ISBN 0–85664–833–7

British Library Cataloguing in Publication Data

Perry, Richard
  Highland wildlife.
  1. Zoology – Scotland – Highlands of Scotland
  I. Title
  591.9'4115   QL259

  ISBN 0–85664–833–7

Printed in Great Britain by offset lithography by
Billing & Sons Ltd, Guildford, London and Worcester

Contents

ILLUSTRATIONS: ACKNOWLEDGEMENTS

1. Geoffrey Kinns; 2. E.K. Thompson, Aquila Photographics; 3, 4 and
5. Dennis Green, The Royal Society for the Protection of Birds;
6. W.S. Paton, Aquila Photographics; 7. Arthur Gilpin, The Natural
History Photographic Agency; 8. Keri Williams, The Royal Society for
the Protection of Birds; 9. Fritz Pölking, The Royal Society for the
Protection of Birds; 10. Pamela Harrison, The Royal Society for the
Protection of Birds; 11. Eric Hosking, The Royal Society for the
Protection of Birds; 12. Dennis Green, The Royal Society for the
Protection of Birds; 13. Geoffrey Kinns; 14. Stephen Dalton, The
Natural History Photographic Agency; 15 and 16. Stephen Dalton, The
Natural History Photographic Agency.

For MILDRED

Whose happiness was in the Highlands

Preface

IN THIS extensively revised, augmented and updated edition of the book originally entitled *In the High Grampians* I have included new chapters on deer, dragonflies, birds of prey, dippers and some of the summer-resident waders; and I am indebted to the editors of *Wildlife, Countryman, Country Life, Field, Scotsman, Scottish Field* and *Shooting Times* for permission to make use of material that has appeared in their pages.

Introduction

DURING the seventeen years or so that we lived on upper Speyside between the Cairngorms and the Monadhliath our first home was in the old township of Drumguish some two miles east of Kingussie, as the rooks and black-headed gulls flew, and high up on the wooded braes of the marshy Spey meadows between Loch Insh and Ruthven.

Our second home was in the village of Newtonmore, west of the Spey, and nearly three miles upriver from Kingussie.

Thus, from the naturalist's point of view, we were in either case in the heart of a unique faunal region.

From Drumguish, with pinewoods falling away on one side and heather moors rising on the other, it was almost four miles eastwards over the moors to Glen Feshie and the undulating, down-like bulk of the Sgoran and Carn Ban Mor, which, mounting to over 3,000 feet, formed the western bulwark of the Cairngorms.

Southwards, the rough road through the birches and alders of Glen Tromie climbed steadily for more than ten miles into the deer-forest of Gaick, which was encircled by hills towering so steeply from a glen only half a mile wide that one had to crane back one's head in order to see the crests of their bosses, rhomboids and pyramids, scarred with granite-pink rock falls and shelving grey screes. At one season they were swathed in purple-brown heather; at another striped and patched with white troughs and fields of snow.

North-eastwards, the way led along the edge of the moors above Loch Insh to the remnants of the old Forest of Caledon in Rothiemurchus, to Loch an Eilein and Loch Morlich, and the defile of the Lairig Ghru that severed Braeriach from Ben MacDhui and the western Cairngorms from the eastern Cairngorms.

From Newtonmore it was a steep climb up the deep green glen of the Calder into Glen Banchor and the almost untrodden Monadhliath. Eight miles south-west of Newtonmore the luxuriant heather moors of the central Highlands were replaced by the grassy moors of Glen Shirra and the Corrieyairack, whose 2,700-foot summit led over to Loch Ness. More than ten miles south of Newtonmore was the pass of the Drumochter, flanked by its great dotterel hills — Beinn Udlaman, Marcaonach and A'Bhuidheanach — soaring to over 3,000 feet; and it was those hills and the Spey itself that dominated the years in Newtonmore after I had concluded my study of the deer, which I described in *The Watcher and the Red Deer*; just as it had been the moors and pinewoods of Glen Feshie and the Forest of Gaick, together with the high tops of the Cairngorms, that had dominated the years at Drumguish.

1 A Traditional Winter

ON A MORNING in mid-August we awoke in Drumguish to find the lawns white with hoar-frost. Three evenings later, when there was a new moon in a clear sky, the wind set in to the north, and at the end of the week a very sharp frost transformed every blade of grass and sedge into a thick white furze. We surmised that we were in for a severe and early winter.

Hardly was September in before the hills were heavily powdered with snow down to 2,000 feet, and early in November the first snow began to fall on the low ground. Once again there was the magical effect of tiny motes of snow suddenly beginning to float down through the green canopy of the pine trees to the warm tawny-brown matting of pine-needles on the road. What primitive emotion in man was it that caused him ever to be as thrilled as a child by the soft, silent fall of snowflakes? Magical too was the almost instantaneous contrast between the familiar green and brown autumn landscape and, in less than ten minutes, a white winter one, as a driving blizzard, dusting the pine clusters and plastering their tall boles, laid an inch of snow on the road. As the early dusk closed in, with snow still falling and the sky full of it, a satisfying sense of being cut off from the outside world possessed me. We were snug with roof and fire, yet in the heart of Nature, in a microcosmic world of our own.

After that first storm a succession of blizzards brought the black-faced sheep in from the high glens, filing down the long road from the Forest of Gaick, one little lot after another, forty in a string, to the shelter of Tromie's wooded glen. The road's frozen, snowy surface was patterned with their sharp hoof-points and skid-marks.

Morning and evening the cold white Cairngorms lay in calm repose against an inky-blue vault of unfathomable depths of colour, heavy with snow. Cold, aloof, infinitely forbidding in their remote icy splendour they might be, but their peace and stillness bestowed on the spirit a serenity it did not find in the lowlands. By day the white dome of the Sgoran glistened in brilliant sunshine: the smooth dead-white sweep of its contours sharply outlined against an alpine-blue heaven.

Through binoculars I could make out black dots on the snowy breast of Carn Ban — skiers! How could I suffer myself to stay in the dull strath when the sun shone all day on those rose-tinted, white hills?

Whooper swans, come down from the High North with the snow, veered in wedges over the wintry moors, stirring my blood with their wild calling. There was some subtle appeal in the music of wild swans and geese that drew me to the open window year after year, to listen for their faint cries high in the starlit skies.

Lying awake through a long night, I would feel the cold biting ever more deeply into my bones, as the frost intensified; but thirty-six hours later we would be opening all windows for relief against a humid 50° F. It was always thus in the Old Year before the ultimate freeze-up in the New Year, for tempestuous thaws and mild days would intervene between one spell of hard weather and the next; days in the first week of December when a pipistrelle bat would be on the wing in the gloomy avenue beneath the lofty, spiring spruce-firs, and evenings when a robin swayed in ecstasy before mate or rival; days when there was the illusion of spring in the air and I longed for bird song. There were sunny days, too, when in the mellowing light of late afternoon there was a depth of colour in the dark green clusters of the pines and in their varnished red-brown limbs that was not apparent at other seasons. Especially bright at that hour of the westering sun was the orange-brown of fallen pine-needles carpeting every road and path through the woods, and the yellow-brown of the terminal sprays weeping from the larches that fringed the pinewoods.

That would be the scene one evening; but twelve hours later the swift Tromie was beginning to freeze over, and in the afternoon the broad Spey was partially frozen too. Hard white frosts, swift thaws with southerly gales, contrasting warmth, heavy falls of snow — that was the traditional winter of the Grampians. For eighteen hours a blizzard raged from the north, burying the moors beneath a white mantle, laying a level six inches of snow on the roads through the glens, and encrusting the pine branches with a soft 'lichen' of snow three inches thick. The few remaining grouse began to pack, flights of blackcock hurtled south and, as the snow deepened, those moorland finches, the rosy-breasted twites appeared. Such was the snow's depth on the high tops that not an outcrop of rock nor vent of spring marred their immaculate whiteness. Marvellous colourings played upon the Monadhliath. The sun transmuted them into mountains of gold ore, but when a cloud passed across the sky they were softened to a pale smoke-blue. At sunset their pearly crests glowed a frosty red. The old hushed loveliness of Christmas lay upon the strath.

Nevertheless, the Old Year went out with fresh, damp, mild days; the grouse packs broke up and squirrels were abroad in the woods. But on the third day of the New Year the real winter set in. Blizzard followed blizzard, while robins, with snowy fronts shining like white stars, were in rivalry. Partidges 'jarred' icily at dusk, grouse packed again, and forty stags came down to the inbye deer-fence. Within a week the drifts of snow on those stretches of the road to Kingussie exposed to the north were several feet deep and too broad and high for the snowplough to break through; and a wreath of snow reaching to the roof of an adjoining outhouse enabled me to ski freely in and out of the garden over the buried fence, instead of having to manoeuvre through the wicket-gate.

Two violent thaws, which bared large black patches on the smooth white tops, also cleared some of the fields and uncovered the warm, deep green moss carpeting the pinewoods, setting the great tits 'saw-sharpening' thinly but persistently, before the storm began again. All roads were blocked once more, and when skiing into Kingussie for rations and mail I was at one minute on the stone-dyke, high above the road, and at the next negotiating a narrow cutting through an opened drift, from which huge chunks of snow had been thrown up on either side above my head.

Skiing with soft susurrus under the alders through the silent white glen of Tromie, I saw only a grouse under a birch tree at the edge of the road and a flock of more than a hundred redpolls and siskins in an alder. But for a full month after that Glen Tromie was deserted by small birds. Three times I skied that way and observed no birds except two grey goosanders.

The blizzards were followed by a frost so intense that the pinewoods lost their contrasting blackness beneath a thick powdering of hoar-frost, and only the rich brown birchwoods coloured the white landscape, until they too were obscured by a greyish-white veil. Thus far the frost had been severe, but not phenomenal. Then on the twenty-third day of the freeze-up, after twenty-nine degrees of frost during the night, an ominous yellowish-grey haze hung in the strath from daybreak to dusk, and during the following night the temperature fell to minus 10°. Eggs froze in their shells, milk solidified in the jug, and beer bottles shattered in the heated kitchen. In the early morning the frigid air caught at the intake of every breath, and the steel bow of a bush-saw stuck to my gloved hand. Among a succession of house-mice trapped at this time was one long-tailed field mouse, a distinctive little beast with its large eyes and ears, sandy-coloured fur, and an orange spot on its white belly. Though the spring from which we drew our water still flowed, the Tromie bore a nearly compact roof of ice, the less rapid Spey was icebound fifty yards across from bank to bank, and the water-meadows were a frozen fen, on which the swans and their cygnets stood huddled, with their black feet and white bodies reflected in the icy mirror.

The frost was hardly less intense the next night; but with the moon past the full, and a diminution in the radiation associated with superb moonlight nights, which flooded the high tops with a light as bright as day, revealing every precipice in their black-fanged corries, the frost gave a little during the day and a gale blew up. Very little snow fell that night, though the continuing wind swept the ground-snow into big wreaths, blocking the newly opened roads again. However, in the middle of the morning another of those violently sudden thaws irrupted. By evening trees were dark green, fields clearing rapidly, and the tops heavily blackened; sparrows were chirping, the first chaffinch was uttering its monotonous spring note, and a blue tit 'belled'.

By mid-February this fourteen-week winter had perished of its own fury, and Badenoch had been transformed into a true arctic tundra. The moors were a black, brown and white patchwork, backed by white mountains; stretches of frozen road were jammed for hundreds of yards with sheets and blocks of ice, which had been forced up from burns and rivers by the spate of melting snows and which had stripped the waterside alders of their bark; and the sea of waters on the Spey meadows were dotted with ice-floes, piled one on top of another on the embankments, the only ground not submerged.

2 Deer in the Snow

A NORTH-EASTER of hurricane force was blasting a blizzard of grey drift-snow through the funnel of the Drumochter. Head-on to the storm, I could neither open my eyes nor breathe and, with the temperature ten below, I was encased in a chainmail of frozen snow. No sooner had the snowplough bulldozed a tunnel-like cutting than the road drifted in again to a depth of two or three feet, and the thought passed through my mind that, unable to hear or to see before me, I was in some danger of being overrun by the plough on its return passage, for the road was flanked by walls of snow eight or ten feet high.

Yet on a knoll a little way above the road a hundred stags were sitting out the worst of the hurricane, with its demented 80 m.p.h. gusts. Facing in all directions, they were almost whitened into the hill. How could they survive such conditions hour after hour, with few breaks in a storm that continued intermittently for eight days, even though their deceptively smooth winter coats were in fact as coarse and impervious to driven snow as the bristling hide of a wild boar? For that matter, how did the grouse survive, crouching in hundreds, heads to the blast and still seemingly strong and lively, on the snow-covered surface of the frozen river and on nearby hillocks? We always knew, in the hills, when a storm was imminent, for it was preceded by large-scale movements of grouse, streaming in packs of hundreds over the moors and along the ridges flanking the strath.

After the storm it was calm and very beautiful, though still freezing hard. Normally, after a blizzard, the white snowscape was flawed by black crags and rocks and dark patches of bushy old heather; but this early December storm, following unexpectedly on several days of zero temperatures, had buried everything beneath a level fall of thirty-six inches, and all Badenoch was covered by one smooth and unblemished, rolling ivory slab of frozen snow. The Drumochter had been transformed into a fantastic arctic terrain. At its 1,500-foot summit, where the county march lay between the Atholl Sow in Perthshire and the Boar of Badenoch in Inverness-shire, the white sweep of snow mantling the great hills and corries, the pass itself and the sheer slopes were coloured only by the bright blue signboards of the rail halt at Dalnaspidal, and barely scratched by the thin line of fencing-stobs stretching up and away into the leaden grey cloud that shrouded the tops.

It was a scene that defied description by pen or camera, though a Japanese artist working on silk, and inserting the tracery of a birch tree here and there, might perhaps have captured the satiny texture of the snowfields in sunshine. No doubt he would have rejected, as inartistic, the dark blobs of grouse squatting in the powdery snow at the moor edge, and also the dark clumps of stags couched comfortably in snowy hollows on one white slope; but he would surely have included the flickering black and white wings of the flocks of lapwings migrating south-west upriver and over the lower moors.

At three places on the frozen road I disturbed buzzards tearing at the carcases of rabbits. A pair of common gulls were beating up and down the road, and three more were standing on the ice of frozen Loch Ericht, whose pale grey desolation contrasted with the brilliantly white hills. In the far background soft blue-grey shadows filled the corries of Ben Alder's white massif. The common gulls nested on Loch Ericht and were, like the black-headed gulls, accustomed to feed well throughout the summer on the scraps thrown to them by lorry-drivers and tourists enjoying alfresco meals at various points along the Drumochter highway.

The five hundred stags of all ages, which habitually wintered on the Drumochter hills, had been augmented by a further five hundred that had come in from the Forest of Ben Alder. Most of the original stock of stags in that forest had been decimated some winters earlier, for with a vegetation composed mainly of sedge and white-grass, but very little heather — the life preserver — Ben Alder offered poor feed in a hard winter. The variable wintering habits of deer were difficult to understand. With the exception of the odd couple of hinds and yearlings feeding apart from the stags, all the deer wintering in the Drumochter were, for example, stags. Yet in Glen Banchor, twelve miles north-west as the raven flew, there were separate herds of hinds and stags; while in Glen Feshie, twelve miles north-east, stags and staggies, hinds and followers might run together in a single herd several hundred strong.

During the noonday hours, if the hills were not under total snow-cover, the Drumochter stags could be seen foraging in the thick heather that grew rank and long high up among the precipitous screes. As the pall of snow spread ever lower down the mountainsides, so the various herds of stags coalesced, and as early as two o'clock, or even earlier, on a January afternoon they would begin to make their way downhill, strung out in long black files across the white slopes, to those places on the lower moors and inbye sheep pastures where they intended to feed during the night. Though purposeful, this downward movement was leisurely, as the stags sank up to their bellies in snowdrifts when plunging through the burns, and lingered here and there to muzzle deep down in the heather or scrape industriously in the snow with swinging hooves. Peaceable beasts (except during the rut) they foraged together with no more evidence of bad feeling than one occasionally shying away from another; while if two, feeding some yards apart, raised their muzzles with nostrils flaring and lips bared back bullishly from their teeth, no more militant action disturbed their harmony than a tentative and delicate tangling of antler points.

A herd of sixty stags might be accompanied by a pack of forty or fifty grouse, which, while scurrying out of the way of advancing stags, found much to interest them in the patches of moss and lichen from which the snow had been pawed away. From these followers one would hear an occasional brief cackle, though the flights of up to two hundred of their kind streaming over the white wastes of the moors at 40 or 50 m.p.h., and jinking and twirling over the high eagle corries, would be silent in such starving weather. Now and again a covey would alight on the road to peck apathetically at the strewn sand and grit, and one had to brake to avoid running over them.

By the late afternoon, when the Pass was deadly white, and the colossal hills and scored, anvil-shaped ridges merged into the sullen blue-grey, snowladen sky, large companies of stags would be right down on the peat-bogs that flanked the road and railway, and twenty or thirty could be seen pacing restlessly to and fro and up and down a knoll overlooking the rail embankment. They were awaiting the cover of darkness, before daring to step down into a burn and cross beneath a bridge to the more favoured moors on the east side of the road. But by ten o'clock the next morning all the companies would be wending their way up the hill again to their noonday retreat high on the screes, with a black file of as many as a hundred and fifty strung out singly right across the white slope of a hill. Despite the constant traffic through the Pass these Drumochter stags remained extraordinarily shy. Although they were not slaughtered by truck bandits the odd beast was poached here and there, and I had only to step a few yards away from the Landrover before those feeding on the lower moors began to move off.

On the seventh day of the freeze-up I climbed a couple of thousand feet up from Glen Banchor to the 3,000-foot skyline of A'Chailleach in the Monadhliath, passing a charm of goldfinches on the way!

Despite the intense cold I was sweating freely, being ski-laden, before I was halfway up the stiff climb over rock and heather, plunging at every step through a crust of snow, ankle or knee deep; but after two hours I was basking in a hot sun on the windless ridge, which was much warmer than the frost-rimed strath. Eighteen inches of snow covered all the ridges above 3,000 feet: range upon range upon folding range of gleaming white barrens breaking in endless frozen seas, with iceberg peaks lit by the yellow sun, as it swung westwards low above the horizonal belt of frost towards the fanged Sisters of Kintail.

One could listen to the silence in those white solitudes, for the only living creatures to be seen up there were deer: a herd of thirty hinds with three or four stags climbing up one long, steep snowface and going down another, halting at every step to paw at the frozen snow, scraping vainly in search of sedge or lichen; tracks everywhere, however, of foxes and hares, but not a ptarmigan to shatter the silence with a profane belch.

As the sun rounded the western hills a rose glow tinted the corrie'd downs of the white Cairngorms, ten miles eastwards across the pine-forested strath; then deepened to lilac, and only the *sgors* were rosy. Then all were a dead cold white against that ink-blue backdrop of the eastern sky; but sunwards long bars of cloud were still fiery, and it was a while after the sun had set beyond Moidart before the orange west paled to a luminous green, which was the Monadhliath's twilight sky at all seasons of the year.

Contemplating this splendour kept me in the high places over-late. The white globe of a full moon had sailed over the frost-furred cairn on the north-east peak of A'Chailleach an hour earlier. So down the snowy sweep of the ridge I went, skiing smoothly, until after an hour I picked up the track by the burn below the red bothy, and was glad to have the frozen footprints of a keeper as an additional guide. The pale moonlight revealed flocks of black-faced ewes clustering on either side of the moor road home, with great-horned rams rutting restlessly among them.

After a week of unbroken frost the snowy moors above Drumguish glinted with the heliograph brilliants of myriads of winking reflectors, as the fragile flakes of ice that fringed every stem of sedge and bent presented their quivering mirrors to the sun. Against that all-white background I could discern the farthest outbye stags, 2,000 feet up on the ridges, at a distance of two miles, though the corries, hollowed out of the gleaming white domes of the smooth enamelled hills, lay in blue-grey shadow.

With no wind to stir the powder-snow or tinkle the sedges' mirrors, no hum of bees, the moors would have been as silent as the arctic tundras on the high tops had it not been for the musical bell-notes from dancing flights of snow buntings and the soft maternal crooning (once heard, never to be forgotten) of hundreds of hinds talking to their yearling calves on a low hill above the flats of bog-myrtle and sphagnum moss, which were rarely so deeply buried under snow that the deer could not scrape clear large patches, and muzzle down to the mosses and the few green blades of grass.

Hinds, calves, stags and staggies were all mixed up together in one hard-weather herd of four hundred, which included a lame staggie and a 'switch' with a broken foreleg. An eagle, circling above them, had his searching eye on a dead six-pointer, with a bullet-hole in its flank, lying stiff and frozen in a patch of long heather. On sighting me, leaning on my ski-sticks some sixty yards from them, the stags moved off casually, but only for a short distance, when they resumed scraping in the snow for forage or couched down in the hot sun. I knew that if I skied towards them, and they cantered across my front, they would not panic, but would gradually draw to a halt if I disappeared from sight in a fold of the moors. Scent was more meaningful to deer than sight, and when a light breeze got up at my back other stags on the crest of the hill, nearly a quarter of a mile away, instantly raised their heads and trotted off; whereas those in the foreground only a few yards away from me, but at an angle to the breeze, did not stir. Their hearing was acute too. Although the noise made by a man's boots crunching over the snow would not disturb them, because it was a natural sound, they would be instantly on the alert if a wire of the deer-fence twanged seventy-five yards away.

Was there a more majestic animal than a heavy-antlered stag standing at gaze on the crest of a ridge, with the breath from his flared nostrils forming jets of vapour in the cold air as his moist, sensitive nose tested the wind? He was the very spirit, the *genius loci* of hill and forested glen.

At noon a shepherd and his dogs appeared over the skyline, whereupon a group of thirty hinds, with a few stags among them, trotted down the hill towards a broad burn, which was three-quarters frozen, and prepared to cross over to the other side. In doing so they afforded me much amusement. Although one of the stags was in fact the first beast to step on to the ice, it was the hinds and calves who crossed direct without any fuss, except for a protesting chorus of blaring and bleating in the general scrimmage along the bank under the gnarled alders; whereas the remainder of the stags kept trotting and cantering backwards and forwards along the bank, unable to make up their minds at what point to cross, and backing up again more often than not when they did venture on to the ice.

Once across, the hinds — ever mindful of their calves' safety — filed some way up the opposite hillside before they considered it prudent to halt and begin feeding again or allow the calves to suckle, which they would do for several minutes at a time. Hinds, having maternal responsibilities, were much warier than stags. Nevertheless they, like so many female animals, had a weakness that tended to offset their natural caution, for they were much more inquisitive than stags. So long as a strange object — such as myself motionless on the opposite side of a deer-fence — was not obviously dangerous, its fascinating strangeness was more likely to draw a hind and her yearling forward, nervous pace by pace, than to alarm her into retreating.

Deer-fences incorporated some of the disadvantages of rabbit-fences. Instead of keeping deer out of agricultural land they might on the contrary prevent them from escaping from the fields, once they had broken in, back to the hills! There were invariably weak points in a deer-fence. Moreover, after the snow drifted against a fence during a storm what had previously presented an eight-foot leap became a mere step over the top wire. Thus, on one occasion I had come upon a group of ten stags on the inbye side of a deer-fence. While the older beasts among them were pacing comparatively calmly backwards and forwards on a trampled beat alongside the fence, the younger staggies were in a panic, throwing themselves against the wires, which a roebuck might have leapt if hard-pressed, but no stag. One staggie was indeed so intent upon escaping that he followed the long beat to within twenty yards of me (waiting beside the fence) before he became aware of my presence and skewed round wildly.

I had presumed that these unhappy stags would eventually escape back to the hills by whatever gap they had gained entry; but five days later I was surprised to find one quite large ten-pointer still on the wrong side of the fence. He may not of course have been one of the original ten, but he was certainly most unhappy. All morning he ran up and down alongside the fence, attracted by the hundreds of his fellows on the hill above; and by the afternoon he had become very distressed, with tongue hanging out and muzzle foaming. Time and again he would bash his antlers frenziedly against the wires or attempt vainly to leap through them. Possibly some stags broke their legs in that way, though one had only to glance at the dead six-pointer lying beside the fence to recognise that those seemingly slender, delicately stepping legs were in fact surprisingly thick. Heedless of danger, the imprisoned stag would trot up along the fence towards me, shying away when only fifteen yards distant; and a second time to within eighteen feet of me! Yet he had only to extend the length of his beat for a further four hundred yards and he could have leapt back to freedom over a low sheep-fence; but he had probably not broken through at that point, for when I attempted to drive him thence he cantered round in a circle and back to his prison bars again.

At two o'clock one afternoon a roe deer came bounding down the hill from the deer-fence at marvellous full-stretch, covering more air than ground. Springing right through a group of a dozen stags, lazing in the sun and not flaring a nostril at her passing, she couched down on a heather flat. Thereupon I set out to stalk her with the sun behind me and shining directly into her eyes as she sat facing in my direction with her brown head just clearing the heather. Although I was too lazy to stalk up to her properly, but only crouched, and made a fearful din crunching through the frozen slabs of snow overlying the bushy heather — which she heard, to judge by the intermittent pricking of those long diamond-shaped ears fringed with black hairs — she allowed me to approach to within thirty paces of her. Only then did she get to her feet leisurely and, after arching her back sensuously, stand to observe me briefly, before bounding away in a series of magnificent leaps, worthy of an impala, rising three or four feet clear of the ground with every leap.

Roe liked to cud and rest in the afternoon, not standing to cud unless they were uneasy; and no doubt individuals had their favourite basking and resting places. I had indeed disturbed a doe, accompanied by a yearling, sitting on this same heather flat two afternoons earlier. Perhaps she heard me, perhaps she saw the sheep running. At all events, puzzled by my half-seen presence, prone in the snow ten paces from her, she gave vent to a prolonged outburst of nervous barks, though she was not sufficiently alarmed to leave her patch of snow at which she had evidently been scraping. And when she did ultimately move away, she soon stopped to look back and actually circled closer again, feeding quite confidently, though lifting her head from time to time to watch the sheep. However, when the shepherd passed she bounded away to the sheep-fence, which she cleared from a standing leap, though surprisingly hitting the top wire; but the yearling characteristically ran up and down the fence looking for a creep-hole, for in my experience young roe deer never willingly jump fences.

In general, roe were much warier than red deer, raising their heads every other second when feeding; but they shared in lesser degree the latter's fatal propensity for pausing for a second look before moving out of danger. I recalled walking up over broken country to within sixteen yards of a solitary buck without any difficulty, before a grouse — always a spoilsport — chased him up out of his hollow, when, before bounding away, he halted while I 'snapped' him. Like the semi-feral black-faced sheep, roe also appeared to recognise the human form at greater distances and more precisely than red deer, and to evaluate the potential danger of a suspicious object. Amusingly, for example, a solitary doe, couched in the hollow near the bottom of the steep hill over the moor, would follow my progress all the way up the hill (on which she was accustomed to see the occasional shepherd or stalker) merely by turning her head, but without bothering to get up from her comfortable form in the sun; but, once I had disappeared from view, did she settle to doze again? Oh no! By the time that I had circled round and stalked down to her hollow she was no longer there. The unseen object was much more to be feared than the one she could see; and a puzzled doe, who had caught only a fleeting glimpse of me, would bark continuously, while feeding nervously in fits and starts: now sitting down to cud, then on her feet again. On one occasion, when stalking a doe and her yearling, but unable to close with them, I eventually got to my feet just as the doe, following the yearling's example, was about to couch down. However, she spotted my movement and proceeded to utter no fewer than twenty-eight consecutive barks — belches would be a more descriptive term — while 'stilting' around, because although I was standing motionless with the afternoon sun full upon me, the wind was in my favour, and she could not verify her suspicions. Ultimately she walked down the ridge across my front, and provided me with an opportunity to photograph her at a range of fifty yards — much too long for an animal of her size.

When homeward bound, pleasurably downhill, from one of my deer-watching expeditions I skied into a party of eight bucks and does who were rubbing their heads in a leisurely way against the rough boles of some outlying pines on the moor, where coal and crested tits were flitting in continual succession from tree to tree with ceaseless excited call-notes and warbler-like trills. Many roe had been scratching up moss in the pinewood at the edge of the moor. Skiing smoothly between the trees, with the soft piping of bullfinches all around, I caught a glimpse of a buck chasing a yearling; but on those blue alpine days Highland roe passed most of the short hours of daylight out on the moors in family groups — here a buck and a doe, there a pair with two youngsters, or just a doe with her inevitable yearling — feeding among the bog-myrtle in the hollows or on rocky knolls swept clear of snow by the winds. As many as twenty might be scattered over one small moor, and almost every doe would be accompanied by one, two or even three fawns and yearlings. From time to time, especially when alarmed, these family groups would amalgamate and be joined by the more solitary bucks to form small temporary herds of eight or nine head. During the long nights they harboured in the birch woods and pine forest.

When skiing on the moors the hissing of the boards over the snow's brittle crust would cause the beautiful mountain hares to skitter away in ones and twos, a dozen at a time, with the fine-spun fringe of long silky hair on their bellies brushing the snow. They skipped crabwise and paused every few yards to sit up impudently, to see what I was about, before skipping away again. To handle them at midwinter was to marvel at the length and silkiness of their white coats. Fluffed up in their snowy *parkas* at the mouths of old rabbit burrows they appeared twice the size of the grizzled, blue-brown hares of summer. When clad in their full white pelage they tended, deliberately or not, to concentrate on snow-covered parts of the moor; but it could not be claimed that this 'camouflage' — dazzlingly white despite its blue-grey base — was of any significant protective value to any hare in Britain. Even on Highland moors there would only be a few weeks at most in any winter when there was overall snow cover; and though some hares lived on the high tops — with a few leverets being born on the 4,000-foot summits — where snow might lie over considerable areas for some months, the majority inhabited moors with extensive peat-hags and sphagnum-bogs between 1,000 and 2,500 feet, for they preferred a bare terrain of peat-hags and gritty, boulder-strewn flats, sprinkled with reindeer-moss. In those years, about once every decade, when their population exploded, they had the habit of appropriating rabbit burrows. When the snow melted in the spring a moor would appear to be dotted with white stones, which were in fact hares sitting up on their hunkers outside their burrows and conspicuous a mile away against the rust-red heather. Nor were they any less conspicuous when they loped and skittered across a burnt and blackened hillside from one burrow or stone cairn to another, while from time to time, when all the moors were under snow in January, one would appear in an almost full summer pelage of bluish-brown. If colouration held any protective value for the mountain hares then it was in the summer, when their smoke-blue fur blended with the natural greys, greens and browns of the summer moors and hills.

Some mountain hares were to be found well below 1,000 feet, which was the upper limit of extensive cultivation throughout most of the Highlands. At that level they trespassed upon the terrain of the brown hares, which did not venture higher, though I never witnessed any conflict between the two. In Badenoch both were to be found at a bare 800 feet above sea-level on the Spey marshes and rough grazings, to which the mountain hares were perhaps attracted by the false reindeer-moss. On one snowy spring day I flushed a mountain hare from its form in a briar-bush on the river bank; and on another day, in May, when looking for oystercatchers' nests on a stony flat by the river, a mountain hare and a brown hare with her leveret scampered away from the same spot on the marsh.

3 Encounters with Stags

ON A HAZY March morning of blue skies, when a southerly breeze was shaking the fat crimson catkins from the old aspens, a stag was asleep at the entrance to the long narrow pass of Guisachan in the Glen Feshie hills. The only sounds in the corrie in which I was lying were the repeated growling *go-back*, *go-back* of pairs of grouse and, infrequently, the song of a lark or the hum of the year's first queen bilberry-bee searching for a nesting hole. Several minutes passed; but the stag remained motionless. I began to wonder whether the antlers sticking up from the heather were in fact those of a dead stag or even perhaps not antlers at all, but the bare snag of an old pine branch; for a stag's antlers, when viewed from the side, stuck up vertically and looked most unlike antlers. I could not recall ever encountering a stag that slept so soundly . . . But suddenly the breeze veered behind me, and the stag became alive, gathering himself to his feet and stretching away after only the briefest gaze.

On fine spring days the herds of red deer tended to break up from their winter concentrations, and the stags scattered widely over the corries, passes and hill-faces in sleepy companies of up to twenty-five head. At the other end of the Pass one young stag out of a group of three kept me lying in cover for more than an hour, for he barely moved from one place while feeding very industriously, getting his muzzle well down into the tangle of heather. He did, however, interrupt his foraging in order to wallow. Lying down in a narrow peaty runnel, he would dig with his antlers and throw back the sopping black peat to one side or the other; but though he was up and down three times he only succeeded in plastering his belly and the shaggy mane at his throat, in contrast to some other stags who, after opening up a peaty sump right across the stalker's path up Carn Ban, had emerged from their mud-baths a splendid overall coal-black, though they dried off nigger-brown. The Guisachan stag was so engrossed in his pleasurable wallowing that he failed to observe the alarm of one of his companions who, pacing up a knoll into my wind, went off like a bullet. His other companion was a cripple with a

29

cruelly shattered hip, though this disability did not prevent him from stumbling along much faster than I could run when I photographed him in evidence. In the end I also attempted to photograph the young stag at a range of fifty-five yards; but even when I was standing in full view he seemed unable to decide whether to trust to his eyes or to his nose and, confused by the veering breeze, advanced towards me initially, before finally trotting off uphill with a nervous, stilted, high-stepping gait.

All morning, while I was stalking these various stags, a solitary hind had sat watching me from the face of the hill above, without ever bothering to get up. Indeed she allowed me to walk at an angle right across the corrie and pass into dead ground, not rising until I reappeared some seventy-five yards away. Even then she stood gazing at me while I photographed a sitting stag, and only finally made off when I returned into dead ground. This was an unusual encounter, for once the winter storms were over it was rare to find hinds on stag ground, since they, accompanied by the yearlings and some of the younger stags, ranged far out over the higher moors. However, heavy falls of spring snow in the corries brought the older stags down to forage in the juniper thickets on the slopes and bottoms of the glens and in the open pinewoods of what remained of the old forest. There, each spreading tree stood well apart from its no less splendid neighbours, and the cone-littered floor was carpeted with spindly blaeberry, which was a favourite spring food of the stags.

Despite their capercaillie and blackcock, their brown owls and sparrowhawks, crossbills, crested tits and redstarts, these ancient forest 'parks' were silent places in the noon heat of a May day, and their quiet was accentuated by the ceaseless calling of cuckoos on the other side of the river. With so many meadow pipits to act as fosterers cuckoos were more numerous in Badenoch than in any other part of Britain known to me, and five or six in a troop would go guffawing and 'bubbling' through the trees. At dawn in May one would call more than ninety times without a break, and throughout the day, from minute to minute and hour to hour, would continue to call from one or other of the pylons that strode across moor and glen on gigantic stilts. At noon his *cuc-oo* would change to a soft, sleepy *cuc-cuc-oo*: a mellow, rather sad triplet, tolled tranquilly over the township with the deep tone of an angelus. Only at ten o'clock in the evening would he fall silent in the before-dark.

There might be one hundred and thirty stags in small groups among the great old pines, feeding in the boggy 'runs' down the steep hillside, and another sixty or more hidden away in the dense maze of juniper in the Caigeann at the head of the glen, where it narrowed to a canyon only two hundred yards wide, walled in on either side by precipitous screes, which swept down a thousand feet from the overhanging crags to the very banks of the river. The dark majestic grove of pines massed in the Caigeann represented the tip of a long tentacle of native pine forest stretching for fifteen miles from Glen More in Rothiemurchus. It, together with the thickets of juniper and the grassy sweeps, the stupendous ravines scarring crags and screes with pink gashes, and the crags themselves casting black shadows down the screes from the midsummer sun, made this east fork of Glen Feshie the grandest in Badenoch. Only in the wild corries on the high tops of the Cairngorms was the irresistible force of flood-water as impressively evident. From a thin crevice in the frieze of crags a burn would gouge out hundreds of tons of gravel and boulders in its downward plunge, and finally bulldoze a channel fifty yards wide and six feet deep through the pines on the banks of the river, carrying fallen trees headlong in its spate; but within a few hours of the spate's subsiding a mere trickle of water no more than twelve inches wide seeped through the stones in the bed of the channel.

These temporary spates, cataclysmal for only a few hours, and the perpetual seepage of moisture down the steep sunless slopes, favoured a special vegetation, of which the dominant plant was the herb-like wood-sage. In dripping chasms and on the clammy loam among the screes were small clumps and single plants of yellow saxifrage, whose bright yellow petals, exquisitely dusted with orange dots and set off by orange anthers, alternated with sepals of palest green. In such damp places were to be found, too, the pink and green flowers of the mountain sorrel, with its characteristic kidney-shaped leaves; the curious drooping purple and orange cups of the water-avens; a giant form of lady's-mantle; and the graceful oak-fern. Grass of Parnassus grew in the damp birchwoods a mile upstream from the Caigeann, together with the little wood-loosestrife, with its gleaming golden-yellow, pimpernel-like flowers and soft green and reddish-green leaves prostrate on the ground.

On a hot day of young summer the stags would be in and out of the forest at all hours of the day — now, up to feed in the bogs and watercourses of the heather moors above; now, down to cool off in the snow-cold waters of the river, or to crop the new grass on the juniper flats alongside the river. At this season they were plagued by the warble-fly. Their defence against this pest was either to canter down into the green twilight of the forest or to mass together in a high-headed huddle, when the acrid 'fried-bacon' aroma of a score of stags milling around possibly repelled the flies.

Observing, one forenoon, that the stags were seeking refuge in the forest, I took up a strategic position against the broad trunk of a pine and awaited events. Cropping the blaeberry avidly, the stags fed ever closer, this way and that among the trees, until there were a dozen within thirty yards of me. Heedless of danger in their sanctuary, seldom did one even lift its head. They were entirely at ease in this their true home, taking no notice of the stalker's wife calling her hens or the dogs barking in the glen only a hundred yards below. That a man should be among them in the forest was a possibility of which only their infallible noses could inform them; but even though there was not a breath of air, surely the human scent would be strong when the range was a matter of feet. Apparently not, for soon they were ringing me round in a circle: great long-legged shaggy creatures — almost wolf-like without their antlers — until the nearest was only five paces from me and I could have pulled a tail or stroked a hide, had I not been encumbered with a camera; and these were not park deer but wild red deer! Pressed close against the rough, deeply corrugated bark of the old tree I tried not to breathe. But still neither the nearest stags nor those behind them were aware of my presence, despite the fact that one young antlered staggie, standing cudding broadside on to me and just beyond the. nearest stag, appeared to stare reflectively at me with his large black and yellow eye throughout the ten-minute vigil. This was one of those moments of which a naturalist dreamed. I had become a part of the deer's world, with no barrier between us. Nor was it as if I was a frozen statue, for not being prepared for a camera subject at such close quarters, I was obliged to raise my right hand in order to adjust the range-finder, and did not hesitate to turn my head and look at the closest stag on my left side, some ten paces distant.

In these circumstances the problem was — which animal to photograph? The stag on my left side, or that 'wolf' in front of me who almost filled the viewfinder? I decided on the latter, for behind him was the cudding staggie and beyond them four or five others just within the forest at the edge of the sunlit moor. Such a photograph would be unique. And then, at last, a pale beam of sunshine streamed through the trees. With infinite care I raised the camera to eye-level and waited tensely for the big stag, still cropping the blaeberry, to step forward into the pool of sunlight. Minutes seemed to pass, but still he fed obstinately in one place in the shade. And then I heard a sudden commotion among those stags farthest away at the edge of the moor, and before I could press the shutter-release my viewfinder was empty.

That was perhaps my bitterest experience as a naturalist. What had alarmed those farthest away so suddenly? Had the camera glinted in a ray of sunlight? Or had a zephyr of a breeze passed from me to them?

Although the alarm had been given, the nature of the danger was not known, and after the initial fifty-yard stampede through the trees, and out on to the moor above, all the stags came to a halt in a ragged bunch, facing in all directions. Then one stepped forward a few paces with enquiring head held high for a better view of the intruder — I had followed them — and uttered a series of deep belching barks as a preliminary to a prolonged stare that lasted for several minutes. Two of his companions also barked, before all wheeled round and pranced away reluctantly; but since they had still not diagnosed me as potentially dangerous and were not very alarmed, they soon returned to the forest to resume their interrupted feed on that choice carpet of blaeberry.

Watching deer at close quarters was easy enough. Photographing them was as correspondingly difficult. Equipped with binoculars only I had enjoyed the not very common experience of having both stags and hinds literally at arm's-length. Red deer, like most wild animals — though not roe deer — had difficulty in determining the precise nature of an object by eye alone. A man lying motionless in the heather, or even on a grassy slope devoid of cover, might be glanced at curiously by grazing deer twenty or thirty yards distant without, however, arousing their alarm so long as they did not catch his wind; even cautious movements might pass unnoticed. Lying within twenty-five yards of three stags feeding slightly below me on one occasion I was able to raise and lower my binoculars slowly as often as I wished without attracting their attention; but when, too cold to lie any longer, I attempted to leave without alarming them I had to sit up very abruptly before they would move, and even then they looked uncertainly at one another as if they could not believe their eyes!

The solution to being able to watch deer at close range lay in first taking up a suitable vantage point and then allowing them to feed up to one. Stalking up to them in the conventional manner seldom achieved satisfactory results from the point of view of either naturalist or photographer, for this almost always caused a slight uneasiness among them and a tendency for them to 'stilt' slowly away, whereas if they were allowed to graze up in their own time they did so unsuspiciously. What was difficult was to discipline oneself to wait. Time and again I rendered a potentially long period of close-range observation abortive by persisting in conventional stalking, when by exercising a little more patience the deer would have come to me, and it was surprising how often they would do this.

Adding camera to binoculars, however, brought a whole train of complications. After some years of experience as a naturalist with deer I knew exactly where I could find them on the hill, in the glen or in the forest on any day of the year, according to the prevailing conditions; though, except in the depths of winter, one had always to go much farther to find the hinds which, more circumspect than the stags, broke away to the most inaccessible moors in their country at the very earliest date that melting snows and the thawing out of frozen ground permitted. Thus, as a naturalist, it was only a matter of taking advantage of reasonable weather conditions and noting the direction of the wind. But, with the addition of a camera to one's equipment, it became exasperatingly obvious that the prevailing south-west wind lay behind the sun for days at a time during the earliest months of the year, so that downwind my scent went with me, while upwind the sun was in the camera's lens. Moreover in winter the brilliant refraction from the snow, rather than providing additional precious light, actually accentuated the unpleasant contrast between white snow and black deer; it also emphasised the fact that in winter the light was not strong enough in any conditions for good shots of deer at the necessarily short exposures that one had to use for animals on the move or likely to move. With stronger lighting during the summer months, the tawny deer more or less merged with their background, especially if one was using monochrome film. Again, if the waiting technique was to be employed, what position did one take up when sun and wind lay together? And while, as we have seen, some movement of the photographer was permitted, the camera was liable to glint in the sun when range or focus was being adjusted.

On an October morning Mac and I followed the drove road to Dalwhinnie and on through the Drumochter to Loch Ericht, snaking away into the misty depths of Corrour. High above the remnants of a hanging larch forest two eagles were sweeping in overlapping, eccentric circles, in and out of the veils of cloud that were coursing along the ridges. From the gloomy southern shore of heather-strewn screes and stunted birch scrub, plunging almost a thousand feet sheer into the loch, a stag roared faintly across the troubled grey waters.

And so along Loch Ericht-side to the vast, desolate sweep of fold and moor, moss and peat-hag about Loch Pattack, and the disconcertingly steep climb to the grim, snow-dusted *circe* of Ben Alder. Yet the folk of township and shieling had dwelt on these moors, and they could indeed be pleasant enough at midsummer when, with light cloud obscuring the sun, the rocky and precipitous hill faces, grassed to their summits, glowed with a vivid luminous green, as if lit by some hidden green fire; while the Cruachan-like mass of Ben Alder and the skyscraping pinnacle ridge of the formidable Hell's Peak were concealed by a peculiar black mist.

Larks were singing over the luscious grassy moors, which could have pastured all the cattle in Scotland, but boasted instead only two white stalking ponies, lonely on a green spit at the edge of the loch. They were curious about unexpected humans in that uninhabited wilderness. There was not a stirk to be seen, nor — thank God — a sheep, nor even a hare. There were no peewits, no oystercatchers, no curlew, no golden plover, hardly a grouse; only a few dunlin, a heron and sandpipers at the loch side; pipits, wheatears, a pied wagtail and a nesting swallow at the bothy far up the Pattack burn and seven miles from the nearest habitation. It was unnatural to hear a robin singing from the dark pinewood above the loch.

Nevertheless, the wilderness was full of life, for it was the season of the Rut, and there were masses of deer, including large numbers of stags, in the pinewoods and on the moors, grazing right up the Pattack burn into the ben itself. We counted five hundred, and many more stags were to be heard roaring from the mist-filled corries. Wallows had been opened up, both in the woods and on the moors, and some of the hinds had been using them. On one low hill more than twenty stags were surrounding what were no doubt mixed harems of seventy-five hinds, though most of the harems we saw were indeed large by Badenoch standards, comprising thirty or more hinds. There was much activity and roaring among those stags rutting, and one big stag even objected to the attempts of a knobber to join his harem of thirty hinds; but though repeatedly chased away, the little fellow eventually managed to double back when the master's attention was distracted by a rival, and join the harem. However, there was as usual no serious fighting among the stags: only the frequent charging out from the harems' precincts of the masters, and the retreating of unsuccessful bachelors. Stags not actively engaged in

Aberarder, and from the heights of Carn Liath and Creag Meaghaidh to the Corrieyairack and Culachy, and so onwards from hill to hill over the glens and corries of the Highlands.

For a long while we stood listening to those primordial voices of the hills; to that strange, stirring, menacing, defiant, lion-like grunting and coughing, lowing and trumpeting, until we could bear the cold no longer and set off down the icy road on the starlit way home.

4 The Rock and its Birds of Prey

THREE miles west of Newtonmore the road to Loch Laggan was dominated by the long triangular 2,350-foot ridge of Craig Dhu and on its south face the immense vertical slabs of the Rock, harbouring Cluny's Cave. The Rock and the two lochans between the birchwood at its base and a partial oxbow of the Spey were my retreat at all seasons of the year.

Only the feral goats passed safely where they pleased across the Rock's precipitous slabs and screes, for though the black-faced sheep strayed on to them from the moors above, some scrambled down on to ledges from which there was no escape, unless rescued by a shepherd swinging down on a rope; while others, less fortunate, fell to the mortuary on the screes below the slabs. The goats were not permanent inhabitants of Craig Dhu, but appeared on the Rock from time to time, after presumably wandering the ten miles across the Monadhliath from the Coignafearn Forest on the upper Findhorn, where there had been a herd of as many as one hundred and fifty for as long as anyone could remember. They were, however, uncharacteristically tame, and one summer day a monstrous black billy, accompanied by a nanny, marched boldly through the village until bundled into a Landrover and hustled back to the Rock. Although spectacular additions to the fauna of the Rock the goats were not really welcome, for plants and seedlings were continually eaten down by the sheep in all accessible places, and further exploitation by the goats on 'inaccessible' ledges was not acceptable.

An ancient ivy 'tree' climbed twenty-five feet up the slabs from the base of the Rock, and the only holly I could recall seeing in Badenoch had sprung from a crack higher up. Many hazels mingled with the gnarled birches among the chaos of fallen rocks and boulders on the screes: typical home of the ubiquitous wren. On a ledge just below the crest of the Rock a dead larch maintained a roothold. Until a winter storm snapped it off at the bottom the larch was the winter perch of a falcon peregrine. In the summer months her whitened eyrie was a little farther west beside the polished black wall of a waterfall; but on any calm sunny afternoon during the winter the falcon could just be descried from the road below motionless on her larch gibbet. I often watched her for an hour at a time, while the sun melted the fangs and organ-pipes of the waterfall's colossal icicles, without detecting any movement, though every few seconds the wintry stillness would be shattered by a loud crack, as ice or frozen rock split, followed by the echoing rumble around the craggy arena of cascading ice or stones.

Since the Rock faced due south it absorbed every ray of the sun's heat, which could be fierce on a windless day of hard frost; and the falcon would sometimes slip away from her larch tree in order to sunbathe in a coign of rock halfway down the slabs. On one such occasion two blackcock and a greyhen came flying over a pinewood, and one blackcock had the temerity to settle on the larch; but it was not long before the falcon became aware of this no doubt unwitting trespass and chased the intruder and its companions for half a mile out over the lochans — but no farther, allowing them to seek refuge in another clump of great spruces and firs on the south side of the Spey. There was indeed quite a traffic of birds between the two woods via the rock, and even a capercaillie might be seen zooming out from the crest of the Rock in a long plane over the lochans.

How seldom did one witness a peregrine make a successful kill. Despite its magnificent powers of flight a peregrine must often be hard-pressed to satisfy its hunger; and it was curious how rarely one saw falcon and tiercel hunting together. I could only recall one instance, when a pair stooped repeatedly, one after the other, striking at a screaming curlew as it twirled round and round in a headlong plunge. They may have killed the curlew, though I could not find any trace of it after the peregrines had landed some distance apart; but I suspect that had they been intent on a kill rather than 'play' the curlew would have been despatched more expeditiously.

It was in the Forest of Gaick that I witnessed a remarkable incident one October day, when the scraggy wood on the precipitous side of the loch was full of migrating passerines — chaffinches, redpolls, bullfinches and redwings, together with scores of wild-flying ringousels whirling from tree to tree with noisy 'chacking' and throaty trills. For some minutes I had been watching six great tits moving south up the glen with slow, slightly undulating flight, forging steadily ahead without dipping down to the trees in the defile, when with a startling rocket-like swish that made me duck my head a blue-grey tiercel (wings bound to sides) stooped past me, to flatten out over the loch in pursuit of another flight of great tits that had ventured beyond the wood. Failing, not surprisingly, to strike its diminutive quarry, the peregrine soared up in a vertical climb, then stooped and missed again; after which it sheered off, while the tits wisely returned to the shelter of the wood. That must indeed have been a hungry peregrine or, more probably, one that could not resist an opportunity to exploit its incomparable mastery of flight and strike. Earlier that same morning I had watched another peregrine harrying more orthodox prey, when a falcon had chased a gaggle of twenty-five pink-footed geese for a mile and more across the moors, before finally singling out one goose which, however, it was able to close for only one abortive strike. Thereafter the goose kept the peregrine at a distance, until the latter ultimately broke off the pursuit.

The aggressive peregrines, no bigger than gnats above the dizzy crags of Gaick and Glen Feshie, waged perpetual war with the resident ravens and eagles. From late summer into winter the eagles and their fully grown eaglets played together over the crags, soaring and plunging, grappling with their talons, occasionally rolling like ravens, though with wings fully spread. If only one eagle was coasting and flapping along a ridge then, as likely as not, a pair of ravens would appear from nowhere to torment it and very quickly drive it down to alight on the side of the hill; after which the two rogues would soar in circles over the glen, congratulating each other with many sepulchral croaks and caws, before making off and disappearing again, while a family of hooded crows protested vigorously.

It had been asserted that eagles were thin on the ground in Badenoch. Yet within twenty miles of Drumguish there were never less than ten occupied eyries in my time. One could not be precise about the hunting area covered by ten pairs of eagles, but one would not be very far wide of the mark if one estimated that in Badenoch this was of the order of five hundred square miles. I would not have described a population of one pair of eagles to every fifty square miles as thin on the ground.

Some of their eyries were extraordinarily accessible. I was far from being a rock climber or even a scrambler; yet one July day I found myself occupying an eyrie, in company with two eaglets. It would be more correct to say that I was standing *on* an eyrie, for the nest had been built on a broad shelf, about seven feet wide, at the base of a craggy outcrop high up the side of a corrie. The eyrie was invisible from below; but as I climbed up the hillside one parent launched itself out from the crag. The nest of heather roots had initially been lined with sheep's wool, but by this time had become littered with the raw remains of rabbits' legs and carcases. When I reached the nest the bigger eaglet, two or three times the size of its fellow, was standing on the luxuriant growth of harebells and lady's-mantle at the extreme edge of the ledge; whereas its wretched little sibling, so small that it was difficult to credit that it was indeed an eaglet, was huddled up against the rock, unable to hold itself upright. When I stepped down on to the ledge the big eaglet appeared to contemplate flight, curving its great wings expressively, but contented itself with hopping down on to a lower slab of rock.

Five weeks earlier, on 3 June, I had located another no less accessible eyrie when, as I entered a glen, an eagle had floated away along the crags, harried by a pair of kestrels, stooping as fiercely as peregrines. This eyrie had been built in a stunted pine tree, adjacent to a dead pine and a third of the way up a steep scree covered with heather and bearberry. The nest, constructed of new pine branches and comparatively small, though deep and bulky, contained a nestling and an egg. Although the nest was about fifteen feet up the tree the scree was so steep that one could look down into it from a range of forty-five feet.

By 12 June the nestling had grown into a colossus with a mantle of white down, a massive hooked beak and large round dark eyes. It occupied the entire nest. It was sufficiently wide awake to raise itself in the nest and loosen its wings when my dogs or I moved around, and was still interested in the dogs a month later, paying much more attention indeed to the yellow collie than it did to me, as it stood hunched up and yawning at the edge of the nest, which had become considerably bulkier from the addition of fresh greenery. No doubt it was surfeited with rabbits, for a whole carcase was lying on the nest. It was still surrounded by dismembered rabbits on 21 July; but eight days later there was no fresh food in the nest, though one parent was present when I arrived in the afternoon, and the eaglet had taken up a position above and away from the nest, affording it a clear view up the glen. It did not move an inch while I watched and walked around the tree, and its fierce gaze never flickered for an instant from its up-glen concentration. On 8 August there was again no food in the nest, and the eaglet was in the same position, still gazing fixedly up the glen. By then it must have been ten or eleven weeks old, and I was not to see it again.

With so many relatively unapproachable nesting sites on ledges and in crevices among the slabs, and in the birchwoods mantling the precipitous slopes, Craig Dhu was a natural stronghold of raptors, which were no doubt also attracted by the variable air-currents on which all might disport. In addition to the peregrines, there were buzzards, kestrels and sparrowhawks, besides ravens and jackdaws and a hoodie or two, while from time to time an inexperienced young eagle would join them. The buzzard tended to be a lazy flier and often appeared somewhat ponderous, though capable of being very agile and stooping swiftly at an acute angle when quartering the ground; but until I watched one harrying a young eagle along the whole length of Craig Dhu I had not appreciated a buzzard's powers of flight. True, the eagle was content to plane, with never a flap of its immensely broad wings. Nevertheless the buzzard, employing that slow, leisurely wing-beat more characteristic of an eagle, played with the latter as it pleased, repeatedly stooping and striking it on the back, and then soaring up like a peregrine; while its mate, away at the far end of the craig, plunged again and again with wings almost closed in victory rolls — as did the male on rejoining her. During courtship the buzzard had a habit of dropping horizontally with extraordinary swiftness before flattening out over or actually through the trees and swooping up again. One thought of the kestrel as a master of hover-flight, but a buzzard could also hover with almost imperceptible wing movement, maintaining position by a constant fanning of its spread tail.

The fascination of the Rock lay in the fact that no sooner did one flier appear, poised on the strong updraught over its crest — whether peregrine, buzzard or raven, or even kestrel or jackdaw — than others would very shortly mount up from nowhere to join in the sport. With a strong wind blowing over the heights as many as eleven buzzards might gradually assemble from various retreats on Craig Dhu and in the pinewoods, and circle out over the lochans, with a peregrine stooping on them, before as gradually dispersing and floating away again to their respective territories. So too, on a day of northerly sleet squalls — ideal for raptor sport — first three and then four buzzards would sail out over the strath from the craig, with a tiercel buffeting one of the buzzards on the back before it had time to roll over. Two of the buzzards gripped talons gently for an instant before mounting to a great height, to poise on the wind while twirling and rocking on silvery wings, talons thrust forward. Then down they and the others dived in twos and threes, to chase each other through the trees. This flighting together was no doubt a frequent occurrence on suitable winter days, but, flighting apart, every Highland buzzard led a mainly solitary existence out of the breeding season. Each had its own territory, and in these well-known territories each was to be found daily throughout the autumn and winter.

Though buzzards were well distributed in Badenoch they were not so noticeably abundant as throughout the west. From Spean Bridge onwards to Fort William and across Loch Linnhe into the long moorland glens and dense birch scrub of the exquisite Morven country, a buzzard crowned a telegraph pole or roadside tree, or stood on an embankment, every few hundred yards; and as one travelled out of the keepered domain of Badenoch and into the derelict west, so there was the pleasure of experiencing once again the fearlessness of buzzards, ravens and hoodies, which hovered inquisitively over one's head. By contrast, when circumnavigating the massif of the Cairngorms and the Forest of Atholl, by way of Tomintoul, Braemar, the Spittal of Glenshee and Pitlochry – a round drive of 165 miles – over endless drear moors and mountain passes on a winter's day, there was, once we had left Speyside, not a single buzzard or eagle, nor even a hoodie. There were just the few pairs of grouse watchful at the roadside, a pack of blackcock flighting over the Balmoral pinewoods and – the only numerous species – kestrels, of which we must have passed more than two dozen hoverers, often in twos. I hesitate to write *pairs*, because although kestrels were commonly distributed throughout the Grampians at all seasons of the year, though predominantly in the straths and over the moors, and very often to be seen hovering over the 3,000-foot ridges, the majority of the extraordinary numbers of immigrants that poured into the Central Highlands in the autumn appeared to be immature birds.

On what did a buzzard prey during the winter, when there were only eight or nine hours of daylight, and that period curtailed by gales, snowstorms or heavy rain? During a hard frost one might, as we have seen, come upon a buzzard tearing at the carcase of a rabbit on the highway, but usually there was your solitary buzzard perched in its accustomed tree or flapping heavily away from it, at whatever hour of the day one passed its particular haunt. How kestrels managed to survive during the winter was equally puzzling. They had always impressed me as birds that had to work remarkably hard for their living. I did not know how many thousands I had watched at their perpetual hovering, but very few had been seen to profit from that method of hunting, in which they swooped down time and again but seldom apparently made a capture. Perhaps kestrels now hovered more from habit than in order to obtain food. What quantity of daily food did a bird of prey require to maintain full health in the wild state? Would a vole or two and a few beetles and earthworms — the produce from a hundred hover-stations — satisfy a kestrel for one day? Perhaps they would. However that may be, as winter strengthened its grip, so every high pass in Badenoch harboured one or two kestrels. With the freeze-up in the Drumochter the only sounds in the cold white silence would be the rushing waters of a high burn, the croaking of one of a pair of ravens perched on a rail, and the shrill *kee-kee-kee* of a kestrel stooping again and again on a noiselessly twirling short-eared owl; for kestrels were as intolerant as other raptors of intruders in their nesting or hunting territories, and when a pair of kestrels nested in the vicinity of a glen eagle's eyrie these diminutive falcons harried their gigantic neighbours as persistently and fearlessly as a pair of peregrines would do. Out of the breeding season it was the unfortunate buzzard who bore the brunt of these waspish kestrel attacks, since the hunting zones of the two were likely to coincide.

Kestrels were constantly hovering over the snow-covered bogs in the Drumochter for eighty or ninety seconds at each station, but only rarely was one to be observed tearing at a small carcase with its tiny hooked beak while perched on a fencing stob. Although neither voles nor mice were noticeably plentiful during the Highland winter, and beetles certainly were not, one presumed that voles did in fact inhabit the long heather and tussocks in the bogs, and were able to survive the extremely severe winters, when snow might lie for more than one hundred days. There was no other prey that might have attracted the kestrels. There were, for instance, no small passerines, though a flock of two hundred snow buntings wintered in a field at the northern approaches to the Pass, as other flocks did at the entrance to Glen Tromie and on other cultivated lands. At any hour of the day, week in week out, even when a savage gale was raging, this flock of buntings could be watched flighting persistently round and round the field — often with one white-winged adult cock at a considerable distance from the remainder — intermittently changing direction and swooping as one bird down to ground level, but without ever alighting; and as they flighted, so they maintained an incessant harsh, insect-like twittering. In coastal fields I had observed very much larger flocks of two thousand or even five thousand buntings behaving in this peculiar manner. Their communal dancing, gnat-like flight, their speed down wind, and their simultaneous direction-changing in a gale of wind were alike remarkable. So few pairs of snow buntings nested in Scotland that one tended to think of them as very solitary birds; but on their Scandinavian and Arctic breeding grounds, from which those wintering in Britain came, many nested in colonies and therefore had some social bonds. Their winter flighting might thus be considered as a communal expression of wellbeing, since, being seed-eaters, only rarely would the blanket of snow be so deep that no food could be obtained.

Although an occasional family of snow buntings might be encountered on the high tops — though not in the nesting corries — of the Cairngorms as late as the first week of October it was November before one heard the soft, elusive call-notes of the winter immigrants on the lower moors or in the glens; and these were very often solitary buntings that came in with blizzards and ran over the snow pecking at protruding herbage. As there were always some buntings on the high tops during the winter it was difficult to determine whether the winter build-up of flocks on the low ground was composed of newly arrived immigrants or of birds driven down from the tops by bad weather. In the strath they fed tamely on oat-stacks in the fields with finches and larks; but most of them had gone by the middle of March, and all had done so by the first week in April. On the high tops there appeared to be a hiatus between the departure of the winter flocks and the return of the few nesting pairs.

It was in the high passes too that the first few short-eared owls — also immigrants from northern Europe — arrived with pro-longed hard weather in the New Year, to hunt for a few days or weeks in the most inhospitable country below the high tops. For three successive Januaries three of these owls hunted over the same moor north of the Drumochter, despite the fact that this moor might be covered with snow, while other moors only a mile or two distant were clear. Oddly enough, since there was no evidence that they nested, up to three owls might also be found on this moor again in late June or in August. As in the case of the buzzards and kestrels one could only speculate as to what they preyed on during the winter, particularly after the rabbit population had been drastically reduced by myxo-matosis. For an hour at a time one could watch the three circling erratically round and across their beat — some three-quarters of a mile in diameter — over bog and moor; and though now and again one might draw itself up, and hover for a second or two on delicately fanning wings, or pounce down with a sudden graceful twirl of its immensely long though rounded wings, I never saw one capture any prey.

Although they apparently arrived at the same time in midwinter and quartered the same beat, each owl hunted alone. But every now and again two would meet and spar briefly, with one twirling up at the other; and a soft snoring cry would sound over the silent moor. After such an encounter one of the two might display, closing its wings and tumbling like a swift, though the manoeuvre was so swiftly executed that it was impossible to determine precisely what happened, other than that the owl dropped and, after tumbling, beat its spread wings so that their rounded tips clashed beneath its body. Only the haunting vocal accompaniment of the cock owl's display-flight high above his nesting territory was lacking.

From time to time one would waft lightly down to perch on a tump or fencing stob or the bare side of a hillock on which the heather had been burned the previous spring; but though it might stand there, with tiny ear-tufts erected, for several minutes, swivelling its maned sphinx-like head to one side or the other over its marbled and mottled back, it was ever on the alert, cocking one eye at a wedge of softly bugling whooper swans forging overhead. And when it ultimately took wing from its perch it would fly menacingly head-on towards me with powerful wing-beats, like a great skua on its nesting grounds; but with this difference, that instead of pressing home a vicious attack, it lifted away in slight alarm at the last instant. In its winter plumage a short-eared owl was almost as white on the underparts as a snowy owl, and when it looked down at me curiously its staring yellow eyes blazed like jewels.

5 Ravens and Jackdaws on the Rock

ON THE first fine frosty day of the autumn a flock of as many as sixty jackdaws would visit the Rock in the early morning for twenty minutes before moving off to their foraging grounds. They visited it again in the early afternoon, when returning from the fields to the rookery in which they roosted, six miles from the Rock. With deep snow down to the 2,000-foot level their presence at the slabs, where some seventy pairs would nest in niches and crevices the following spring, was somewhat incongruous in October; but these sporadic late autumn visits from the first week in October until mid-November — and then not again until late in January — in order to scrap and play around the nesting cliffs, which they had deserted in August, were a feature of jackdaw life in Badenoch.

The inimitable Konrad Lorenz might have explored the life history of German jackdaws down to the last detail, but a fascinating study of Highland jackdaws awaited any naturalist in search of a project. In Badenoch, for example, there appeared to be four discrete populations of jackdaws. First, there were the largish colonies of up to a hundred pairs nesting in crags or screes up to a height of almost 2,000 feet, or in the ruins of old castles or barracks. Then there were those that nested as individual pairs in groves of old birch trees. The third category comprised the few pairs that nested in rookeries. And finally there were those pairs that nested on houses in the villages, and these were numerous. Although these village jackdaws flocked freely when foraging in the gardens, and took part in communal flights about the village at frequent intervals, they remained in pairs all the year round, and my favourite distraction when shaving in the morning was watching the males affectionately nibbling their mates' heads on the chimney-stacks — an attention which was appreciated for limited periods only. However, the young ones disappeared from the village as soon as they were fully fledged, and presumably joined the widely foraging rookery jackdaws, leaving their parents free to titillate on the chimneypots again. All the non-village jackdaws roosted with the rooks during the winter, and on the long flight from their feeding grounds to the rookery one flock would link up with another, until the roosting flight might number as many as 250 birds. No doubt each flock had its own special feeding place, and a pied individual known to me fed in the same field in five successive years; but why did some jackdaw communities forage with rooks, while others did not? Individuals in the roosting flight would call 'derisively' to solitary pairs flying silently and far below them in the opposite direction. The latter were perhaps on migration, for they flew fast and direct (ignoring the calls of the homing birds), swinging from side to side as they veered on their compass bearing.

One October afternoon two ravens came junketing aggressively high over the Spey, hurtling along shoulder to shoulder in swift, purposeful flight, buffeting and digging at one another with their powerful beaks; but soon after reaching Craig Dhu one broke off the engagement and returned across the river. *Kronk*ing with satisfaction as it soared, intermittently plunging and rolling, it ultimately stooped from a great height to the crags whence it had come, a mile distant, and rejoined its mate; but there they, in their turn, were harried by a peregrine. No time to be bored if you had wings and company.

What pugnacious and high-spirited birds ravens were — perpetually at war with peregrines and for ever harrying eagles and buzzards, or merely playing with them, as in one instance when four ravens and five buzzards, soaring together, joined forces to persecute an eagle. They also warred or played with their own kind, and even during the breeding season single ravens and also pairs might be observed speeding with urgent flight to meet up with others for a bout of sparring, to the accompaniment of appropriate vocal comments, before sheering off their separate ways. Such inter-specific combat was particularly common in the autumn when the parents were evicting the young birds from their territories, and again in the New Year, as late as March and even April, when interlopers — possibly the previous year's juveniles — trespassed upon territorial air-space.

Ravens (like raptors) obviously found their powers of flight, and especially perhaps their famous roll, exhilarating. To the latter manifestation of their *joie de vivre* I paid particular attention. Although rolling was not exclusive to the breeding season it did reach a peak of frequency during the first three months of the year, notably in February when the adults were established in their nesting territories but before the young hatched. However, there was a second peak period from August to October when the ravens were constantly engaged in war and play with the raptors; but I never recorded a single instance of a raven rolling in May, June or July except as a defensive measure against attack.

Both cock and hen would roll — the latter less frequently; and when the pair flighted together, first one and then the other would roll, creating the impression that they were rising and falling on fluctuating jets of water; or both might roll while covering great sweeps of moor, first diving with wings closed and then turning over from right to left (and back again the same way), and also dropping ten or twenty feet on their backs, with wings still wrapped tightly around their stiffened bodies, while *kronk*ing with open beaks. Their exhibition would conclude with a couple of somersaults while falling seventy-five feet.

The unique feature of the roll was the leisurely ease with which it was executed. One could not but grin, no matter how often one had seen it before — whether it was a cock launching out from a stunted pine on the moors to roll only a few feet above the heather while in pursuit of its mate; or one of a trio sweeping around a crag, slowly and deliberately closing its wings and turning smoothly left over on to its back, stiffening head and body as it did so, and then turning back the same way. It might repeat this slow roll several times, but always turn over left about. Variations of the roll included turning over with wings fully outstretched, as an eagle would do, or dropping a hundred feet on the back with 'wrapped' wings.

Although the roll and the steep, shut-winged plunge might be associated with a variety of vocal accompaniments — the familiar *kronk-kronk*, a harsh caw, a 'woody' *koop-koop* or *kyowp* — there was one that was seldom heard except in conjunction with the roll; and an extraordinary utterance it was: a batrachian *psee-ong*, with the *ong* vibrating like an electric gong for several seconds. Though not a loud note it carried a great distance.

About a week before the ravens had eggs, rivalry between them on the crags of Craig Dhu and the buzzards in the woods on the slopes became intense. All day long the ravens planed and soared to and fro between crag and woods, repeatedly rolling, while 'playing' with the buzzards. As yet there was no venom in their play, but a joyous abandon in their powers of flight, with a buzzard falling with rocking wings and downstretched talons, and a raven corkscrewing down and rolling when beneath the buzzard, so that their spread wings appeared to clash like cymbals, while high above them soared the tiny figure of a crescent-winged sparrowhawk. Then raven and buzzard plunged below the level of the trees and pursued each other through them. Later, the two ravens sailed out on separate courses half a mile apart. In the cold bright stillness of a frosty morning I could hear the silken lash of glossy wings when the cock opened his after somersaulting.

As soon as the hen began sitting on eggs the cock, stationed on his customary lookout crag, would utter a booming 'oaken' *kwowoch* — which was the most impressive of all his numerous calls — or a high-pitched though mellow *cruk* or *crerk*, which was perhaps an affectionate rendering of his everyday *kronk* and a reminder to his mate that he was present and on guard. At this stage he was extremely wary, shifting from crag to crag when I approached and snarling rather than *kronk*ing; and his anxiety led to incessant sparring and ten-minute engagements with the buzzards, to the accompaniment of continuous snarling and mewing. If the hen raven mounted to join her mate, then the pair of buzzards poised above them with talons dropped. If the ravens outmanoeuvred and stooped on the buzzards, then the latter in their turn rolled over on their backs, but with wings spread and talons outstretched in defence. Both ravens might then quarter the buzzards through the woods, with the hen chasing them right past the nesting crag, while the cock stationed himself on his lookout crag once again.

In the days before the young ravens hatched the hen brooded them closely and played no more with her mate, who roosted in a niche some thirty yards from the nest. He, however, still occasionally mounted to a height from his lookout crag, to indulge in his now waning fervour for rolling, when he might somersault three times in succession. And though the hen would leave the nest when I climbed up towards it, and beat swiftly to and fro, uttering the repetitive *cog-cog-cog* she employed when driving away the buzzards, and then a *pruk-pruk* when she had flown farther away, she would drop quickly on to the nest again as soon as I was clear of the crag, while the cock kept up his warning *wonk-wonk*. The infernal squawking of the young ravens lured me to a most unpleasant scramble up the wet and rotten slabs. But when I finally reached their crag I could not locate the nest, nor could I climb down again; so was obliged to continue on up to the summit of the Rock — in a cold sweat, and without even a creeping spray of purple saxifrage as reward, for this beautiful saxifrage with its very large magenta flowers was sparsely distributed in Badenoch, and I had found it only between 2,000 and 3,000 feet in Glen Feshie, though it had also been reported from the shingle beds of the Truim at the entrance to the Drumochter.

The fledging of the young ravens was the cause of much excitement to their parents. With the massiveness of their heads and beaks accentuated by the shaggy feathering of their deeply pouched throats the adults were immensely impressive, whereas by their side the young, with their sharp-cut heads and bodies, glossed brown or purple, could have been crows, except for their curved beaks. Though fairly strong in flight the fledgling ravens were initially a little uncertain in their landings; but after they had been on the wing for a couple of weeks the family became a constant nuisance to all their neighbours. Within six weeks of fledging a young one might be found feeding alone on a dead sheep on the screes. *Wa-aa* it would croak harshly on espying me, though when answering a distant parent its croak would change to a high falsetto. Young Highland ravens were fearless, and on a number of occasions, especially when accompanied by my collies, I had them hovering inquisitively over my head. The dogs they would mob with downstretched talons for several minutes, while *kronk*ing and snarling and also uttering a variety of other calls, including thin whistles, wheezing squeaks and an extraordinary gurgling *gutta-chook* that could perhaps be likened to the sound of a cork being drawn from a bottle.

6 Summer and Winter Visitors to Badenoch

IT WAS in the last week of March or early in April that ring-ousels returned to nest on the Rock. Since one never saw them migrating through the glens and passes in the spring as one did in the autumn they presumably flew direct to their breeding localities, in which they apparently arrived in pairs. Thus one day on the Rock, or in some high glen which had been songless the previous day, I would hear once again that thin, fitful, thrushlike piping; and curiously enough it was only in those early days that one might hear — at twilight on a misty evening perhaps — what could be described as the ring-ousel's true song: a brilliant musical warbler-like trill or yodel. This would be preceded either by a hard *chack* or throaty chuckle, or by the ringing *phee-tew*, *phee-tew* of a thrush, or by a clear, sweet single pipe, repeated three or four times at intervals, and powerful enough to carry a quarter of a mile above the roar of a waterfall, though with a veering breeze it was barely audible at half that distance. But the most characteristic song was that endlessly reiterated, soft and mournful whistle, whose sad refrain echoed all day in one's consciousness. Three cocks might sing against each other with wild whistles from their various song-stations on rock or pine tree or towering pylon, and where they nested in colonies their song continued from dawn to dusk — though very faint in the afternoon — until the end of June.

The largest colony of ring-ousels in Badenoch was in Glen Tromie. There, at noon on a July day, as many as thirty might be collecting food on an old township's narrow strip of greensward and thickets of juniper beside the river and below the steep hillside. Since the bulk of the juniper berries did not ripen until long after the ring-ousels had emigrated in the late summer they probably did not feed on them, but nested in the thickets, which attracted many kinds of birds; and where there were dense clumps of juniper, there you could be certain of finding colonies of ring-ousels, whether the glen was mainly wooded with birch and alder as in the case of Glen Tromie, or with pine forest like Glen Feshie. But ring-ousels, like so many Highland birds, had to be sought for years before one began to understand their habitat preferences and distribution, and could form an approximate estimate of their numbers, which varied considerably from one year to another. In Badenoch I recorded about thirty breeding localities over an area of some three hundred square miles, and found that while ring-ousels could be classed as locally numerous up to an altitude of 3,000 feet, the majority nested below 1,500 feet, and large colonies were never established very far from the floor of a glen. Were I asked to describe their typical habitat, where junipers were not present, I would say — steep, treeless slopes of heather, crags and scree flanking the higher glens, and precipitous ravines with here and there, leaning out over the foaming falls of a burn, a stunted rowan or birch. On one memorable May morning in the Forest of Gaick, for instance, I enjoyed the marvellous experience of listening to fifteen cocks singing along the three-mile length of the glen.

In such country the stranger would have a very fair chance of finding at least one pair of ring-ousels between the bottom of a glen and 2,750 feet — at which height I watched two pairs in the Monadhliath one April day chasing about on the glistening snow of a cornice lipping an immense corrie. In the Cairngorms they nested even higher, and I was startled one noonday towards the end of June, when going down into Glen Feshie from the high tops, to hear two or three cocks piping in the deep amphitheatre of Corrie a'Gharbhlaich, where the red deer calved just below the 3,000-foot lip of the corrie. There were always insects on the snowfields that lingered into late summer on the Cairngorm tops, and these possibly attracted ring-ousels, for on 30 July one year there was a cock on a snow-bridge across a ravine. Odd pairs were indeed to be met with in the most unlikely places — beside Loch Einich, for example, or nesting actually in the summit cairn of a 2,000-foot hill; and they were to be found in these remote localities year after year, even in such high and wild passes as the Drumochter, where however, they were sparsely distributed, though formerly reported as numerous. Habitation, with the exception of the loneliest shepherds' cottages, was shunned. Having made these various points, it would be fair to assume that the one habitat in which ring-ousels would not be found would be wooded terrain; yet paradoxically two out of the three largest colonies in Badenoch were in or near extensive woods, presumably because of the presence of considerable 'commons' of juniper.

By the middle of August most breeding stations had been deserted, and early in September flocks of as many as twenty-five ring-ousels were to be met with in such high passes as those that led south from Gaick and Feshie to the desolate moors, peat-hags, bogs and granitic wastes of the Braes of Atholl; and flocks of this strength continued to frequent these passes, especially those wooded parts where rowan berries were abundant, as late in the autumn as the last week in October. More often than not these migrant flocks were composed exclusively of cocks, and were frequently accompanied by redwings, mistle thrushes and song thrushes.

By the middle of April a million small primrose plants would be leafing among the withered bracken, dry and sun-crisped, on the steep slopes at the base of the Rock, with here and there a short-stemmed flower. The earliest brazen celandines gleamed among the flowering dog's-mercury, attracting many tortoiseshell butterflies. Kestrels chased and courted with high-pitched teetering cries from ledge to ledge; then plunged down the face of the cliff with wings closed, or fell in pairs with grappling talons. All the small passerines were in the birchwoods or in the plantings of conifers — song thrushes, black-birds, robins and dunnocks; bullfinches, greenfinches, chaffinches and crossbills; siskins and redpolls; pied wagtails; treecreepers, goldcrests and blue, coal, great and long-tailed tits. In May and June they were joined by redstarts, tree pipits, spotted flycatchers, wood warblers and multitudes of willow warblers, whose sad sweet diminuendos were so pervasive that they prevailed over all other bird songs except those of the chaffinches. Their normal delicate plaints were in marked contrast to the loud and persistent, lesser woodpec-ker-like *pee-pee-pee* uttered by the cock willow warbler when, upstretched on a branch with wings fanning feverishly, he displayed to his mate, who might also beat her wings. Only a circle of white sheep's wool in the bracken below revealed the nest, with its minute entrance hole. With a similar shivering and waving of wings one would betray its anxiety when perched high above a cuckoo in an ash tree, while a pair of tree pipits looked on unaffected.

By May the longing to surrender to the sun's heat and lie listening all day to the birds singing would be overwhelming. In the Rock's rare suntrap in this northern strath, or in the green curve of the Calder's deep glen that baffled the arctic blast from the north-east, one could relax in tolerable warmth for an hour or so, and be roused only by the exotic rivalries of one cock redstart chasing another and of two fighting furiously and tenaciously, both on the ground and in the air, when their quivering flame-coloured tails transformed them into birds of paradise. Its objective achieved, the pursuer or victor would perch in a hazel only two or three feet above the ground and utter its five-note burst of song, which somewhat resembled the chaffinch's delivery. This might, however, be followed at leisurely intervals by a variety of hurried, detached and dissimilar phrases, which were startlingly nightingale-like in their liquid perfection and brilliant execution; but a redstart could seldom be watched closely in song for any length of time, since it usually sang from a well-concealed perch and offered only a glimpse of that flaming tail when it dived across a clearing, though now and again it might utter a phrase from the extreme tip of an alder tree, permitting a brief view of its glistening silver-white brow. Nevertheless, redstarts obtained much of their food on the ground. Wherever there was a birch or conifer wood, there you found the redstart in Badenoch, as high up the glen sides as 1,750 feet, but always at the moor or road edge of pinewoods.

The black redstart I saw only once in Badenoch, and that was an immature cock on 5 November in Glen Tromie, where typically it was perched on the roof of a deserted cottage.

As a southerner born and bred I was never able to accustom myself to being visited by *goldfinches* during the hardest weather of the desperate Cairngorm winter, nor to accepting them as regular harbingers of severe cold like twites, snow buntings and whooper swans. Goldfinches belonged to the sun; to blue skies and orchards of blossom. Yet on one occasion when the Old Year was nearly out a charm of five goldfinches passed leisurely eastwards through Newtonmore, flitting from one clump of ragwort to another of knapweed; an hour later a wedge of five whoopers also forged east; and by the afternoon the first blizzard of the winter was raging in the Drumochter. Again, in November one year, after the sun had set with an ominously smoky glow smouldering over the white craters of Ben Alder, twenty miles to the west, there were ten degrees of frost in the sun-lounge at midnight, and we woke in the morning to windows opaque with twenty-eight degrees of frost. Zero temperatures in November! Not unusual in Strathspey. Yet at first light on such a morning four goldfinches were twitter-singing in the garden.

Goldfinches did not visit Newtonmore every year, though I recorded them in gardens, on waste ground and beside the Spey in every month of the year except June. They appeared most often and most numerously during the winter months from October to March, and especially with very hard weather in November and December, whether this was in the form of a heavy snowfall, severe frost or a blizzard of hurricane strength. Their arrival in these conditions was the more marked because an intervening mild spell might have resulted in the disappearance of every goldfinch in the strath; only for them to reappear with the next cold spell.

In a favourable year the typical pattern was for single goldfinches or pairs to pass eastwards or northwards through Newtonmore during the months of July, August, September and October. These travellers might be heard in full, though soft, song both in flight and when perched in a riverside alder or on the top of a garden spruce. Then, with very cold weather towards the end of November a build-up would begin, culminating in charms of as many as a dozen feeding on beds of knapweed under the snowy birches. They might remain in numbers until the New Year; but after that only odd birds (often in song) were to be seen passing until May.

The origins of these goldfinches, and also the circumstances in which they visited and possibly wintered in such an untypical environment, were obscure. However, a few pairs nested in Perthshire, and I concluded that they were British natives rather than Continental immigrants.

Goldfinches were not our only hard-weather visitors. There was, for instance, that musical buzzing note when we were having an early cup of tea one November afternoon. It took perhaps less than half a second for my memory cells to register, evaluate and correlate this sound with a snowy afternoon in the alder woods of Glen Tromie a year or two earlier. We were accustomed to have crossbills lined up along our roof gutters, but to have a *waxwing* on one and peering in at the window! — that was a delight I did not expect ever to experience again. This bird was one of five in the garden — after a spell of hard weather — engaged in stripping a cotoneaster bush of its fruits, and the brilliant red berries were no brighter than the curious 'wax' tips on their wings.

Few winters passed without our seeing at least one flock of up to twenty waxwings. They were usually in brier bushes on the low ground, but sometimes in the higher glens or even flighting over the moors with their unmistakable sibilant trilling which, when they became aware that they were being observed, would change to a prolonged bell-like alarm call, for waxwings were not silent birds as had often been alleged.

Living in the exact centre of Scotland, we could not expect many immigrant rarities. Thus any stranger in the strath was particularly welcome, affording me pleasure and speculation about its origins for several days subsequently. Almost every year an exotic visitor from northern forests wintered from late in October until early May on that part of the moor above Drumguish where small outlying pines had seeded from the forest of Badan Dhu on the hill above. This was a great grey shrike: a monstrous black and white striped 'chaffinch' with ivory-coloured underparts. How extraordinary that a single shrike should migrate all those hundreds of miles in order to winter alone on this heather moor many miles from any other of its kind! The tips of the low, gnarled pines served as convenient lookout perches from which it could drop down into the long heather from time to time to pick up some small prey. It would also hunt by hovering — aligned on a vertical axis from huge hook-beaked head to the tip of the exceedingly long diamond-shaped tail, which was continually dipped and flirted — before pouncing down. During the winter it presumably subsisted on beetles and other invertebrates, for I never saw it capture a vole or chase a small bird.

Goldfinches, waxwings, shrikes and the resident crossbills — these provided the highlights of the mainly dark and drear Old Year, though once in a while the weak glow of a late afternoon sun would light up a rich red hill of bracken, flecked with gold specks of sheep wending their way up to their night places.

Migrations through Strathspey involving large numbers of birds were restricted to thrushes, though a few hundred geese passed in most years. The latter consisted mainly of greylags and pinkfeet, but occasionally of whitefronts. The geese did not observe the normal migratory rules of following fly-lines along rivers or straths or hill ridges, but travelled on a fixed bearing from north-west to south-east, crabbing resolutely across an unfavourable wind when necessary, and breasting the 4,000-foot summits in their course.

In the early days of November every rock and boulder on the moors would be splashed with the dark purple or magenta-coloured droppings of berry-eating fieldfares and redwings, migrating south and west in large flocks. At that date they, together with fewer blackbirds, mistle thrushes and song thrushes, would have been passing in immense numbers for three weeks. The bulk of the passage occurred in dull stormy weather with strong winds, when the cloud base lay as low on the hills as 2,000 or 3,000 feet. In such conditions many thousands of fieldfares and redwings migrated through the strath every day, with most flying at heights of less than a hundred feet. Mixed flocks might contain as many as two thousand birds, with fieldfares in the majority, but discrete flocks rarely exceeded one thousand fieldfares or five hundred redwings. Their migration continued from dawn to dusk, though the largest concentrations passed through before mid-morning, and flocks might still be mounting up from resting places in the trees at the darkening or at moonrise. On clear days, however, there might be no passage at all, while when very low cloud and mist resulted in a hold-up, hundreds of redwings could be heard twittering and whistling softly from a grove of larches.

Their migratory movements extended over all Badenoch on a broad front and included all those glens that gave access to passes through the hills, with the result that as many birds might migrate through such high glens as Gaick as through the main strath. Small flocks, especially of fieldfares, were often to be seen flying along crags and hillsides at heights of up to 3,000 feet, alighting now and again to feed on bearberries and bilberries. Sometimes these high-flying fieldfares would mob an eagle, and be harried in their turn by hoodies and ravens. In one incident indeed a raven caught a fieldfare on the wing, though subsequently dropping it.

Where the strath widened to a breadth of a mile or more the fly-line appeared to be from east to west, as the migrating flocks swung across from one wood or belt of trees to the next. Woods invariably attracted them, and when the clouds were low they would sweep down to the trees in scores or hundreds to rest a while before journeying on. However, once the western barrier of the Monadhliath had been reached, most of the flocks, though not all, appeared to swing southwards in order to travel south-east through the Drumochter into Perthshire, rather than to continue on westwards by way of Loch Laggan.

In stormy weather or blizzards there was much doubling back north or east of small flocks of both fieldfares and redwings, together with confused movement in all directions. In contrast to their normal leisurely, though steady progress from wood to wood, these erratic movements were more hurried. Small flocks of fieldfares in particular would suddenly appear from nowhere, corkscrewing down with extraordinary velocity to alight on the trees, drawing attention to their presence by their curiously squeaky call-notes, before setting off again in the reverse direction.

Redwings and fieldfares no longer wintered in Badenoch, and the majority would normally have passed through by the middle of December at the latest. Exceptionally, however, the passage of fieldfares would continue with barely a day's cessation from the middle of October until the last days of January.

7 Crossbills in Larch Grove and Pinewood

THE COLOURFUL plumage, engaging habits and incessant chatter of the Scottish race of parrot-crossbills atoned in some measure for the almost total absence of birdsong for more than half the year — one's most serious aesthetic deprivation in the Highlands. Even on the greyest of dull winter days the rose-red cock crossbill conjured up for me the hot sun of Charente or Champagne. His apple-red colouring was unique among British birds. His mate was apple-green, with apple-yellow rump and flanks, just as in the cock these were an even brighter shade of that rich, warm roseate of his mantle.

In Badenoch the crossbills' staple food, except during the breeding season, was the seeds of larch cones; and it was in the larch groves that their flocks were most often to be found. There were a number of these groves within seven miles of Newtonmore and also around Drumguish, and since in either case it was my good fortune that the nearest were within a few hundred yards of my house, there were few weeks most years when I did not see crossbills. Moreover, in those occasional years when the crop of larch cones partially failed, the seeds of pine cones, and no doubt of spruce also, were the main standby. At such times the quarter-acre of pine trees at the back of my house in Newtonmore apparently provided an inexhaustible supply of food, despite the fact that the crossbills' method of feeding appeared to be excessively wasteful, with multitudes of cones being discarded and dropped prematurely or even thrown down without a single seed being removed. Yet a single spruce cone might be worked at for ten minutes or even half an hour.

In some years the flocks of crossbills intending to winter in Badenoch were already at full strength by July, but it was usually in the last days of September or the opening days of October that the first wintering flock of eight or nine would announce their arrival by flighting noisily high over Drumguish and showering down to settle on the spire of a tall larch: there to commence feeding on the cones, to the accompaniment of a continuous chirruping chorus. In some seasons they frequented the Drumguish larch groves throughout the winter: in others they visited them only sporadically; but if they were going to winter in them permanently then very soon after the arrival of the first small flock their numbers would have increased to thirty or thirty-five, and they were to be found almost daily feeding high up in the larches in company with coal tits, goldcrests, redpolls, chaffinches and perhaps an irascible squirrel with angrily lashing tail. Hanging upside down, like little parrots, from the supple twig-ends, they wrenched off the small tawny cones very easily and neatly by their stems and carried them in their beaks to nearby branches. There, standing on them with one or both feet, they would gouge out a seed here and another there with twisted lower mandible. The woody impact of cones falling on hard branches and their dull thuds on the ground beneath would be continuous. Occasionally a cock would present a severed cone to its mate. But though all were constantly at work their mellow chirping *tyup-tyup-tyup* seldom ceased for an instant.

Once in a while a pink cock would perch on the extreme tip of a larch, or on the top of a pine or spruce or ash tree, and utter its canary-like song, which was based on a reiteration of the choral *tyup-tyup-tyup*, rendered jerkily and squeakily like a linnet; but the songs of bullfinch, nightingale and siskin all offered points of comparison, though particularly that of the bullfinch, with its high-pitched piping notes. Short stanzas might be of quite brilliant siskin quality, while rich throaty trills, a *jug-jug*, and a strong whistling *tü* belonged to the nightingale. Although brief, fragmentary song might be heard on any day from the beginning of July until early the following June, it was especially fine and frequent in October and November; but true sustained song from one cock in answer to another with that clear six- or seven-note *pee* and its *sotto voce* accompaniment, or with the mellow lower-pitched six-note *tew*, was uncommon enough to attract marked attention. Occasionally a cock would circle out in song-flight from a tree at the edge of the grove, or from treetop to treetop. With cruciform wings beating slowly and tremulously, this joy-flight resembled those of greenfinch and siskin, and was associated with the nightingale notes and other short though brilliant musical inconsequentials which it was impossible to transcribe. Sometimes a hen would sing a few notes, and both the adult red cocks and a few of the immature cocks (with only a pale red tinting on their breasts) continued to sing in tinny chorus right through the January frosts, and were still singing in the middle of February when the flocks began to break up. That event put an end to choral singing, though small bachelor parties of adult cocks and the golden immatures maintained their incessant chirping.

In the late afternoon sun of short winter days seven or eight crossbills, and sometimes the whole flock, would gather almost daily in the great ash tree that overhung our garden at Drumguish — though less frequently during snowstorms — and a hen with shivering wings might invite a cock to neb bills with her. He would respond by presenting her with a cone or would regurgitate a milky fluid into her beak, staining her breast with the overspill. If crossbills did not mate for life, then pairing probably took place during the late autumn and early winter, since the courtship flight was first noticeable in October, while from early November until February pairs were to be seen nebbing affectionately. From time to time two or three or more would drop down to drink from the roof gutters, and it was a fine sight to see ten red cocks lined up along one. Crossbills were thirsty birds — not surprisingly so, having regard to their dehydrated diet of conifer seeds — and would often fly down to a rain-pool or to the banks of a burn. In contrast to their usual restlessness in the trees they were strangely confident on the ground, and only when watching one drinking ten feet away was it really possible to distinguish the twisted tips of the powerfully arching bill. And then they would be up and away back to the trees with a burst of rich song-notes.

By December the various flocks were at their maximum strength of between forty and sixty birds, with the red cocks outnumbering the hens. Then from January to mid-March the flocks decreased in the size again with the breakaway of breeding pairs, though as late as the end of April snowstorms would bring flocks of as many as forty, presumably non-breeding birds, into the larch groves again. By contrast, blizzards in January and February were likely to result in the total disappearance of all crossbills from the strath. Since their food supply remained both intact and accessible I could not think of any reason for this exodus. It was true that in such conditions I would find small flocks in the old pine forests in the high glens, but the total number of birds in these never approximated to those that usually wintered in the strath. Every year I would ask myself whether the large wintering flocks in Badenoch were not Continental immigrants rather than Scottish natives; but I never once heard a call-note indicating that they might have been the former. It was, incidentally, on 3 August one year that I was brought perhaps the rarest bird that I had ever handled. This was an adult two-barred crossbill. Smaller and more chaffinch-like, with the two white bars on its wings, than the Scottish crossbill, the blood-red mottling on head, shoulders and breast was even ruddier, but the right-hand twist to the lower mandible was much less pronounced. Another specimen was reported from Foula on the same day, and they were no doubt vagrants from some Scandinavian forest or Siberian taiga.

Although wintering flocks in Badenoch originated from the amalgamation of families of adults and young, mature cocks appeared to be in the majority from October to February, some small flocks were indeed composed exclusively of cocks. There were days when they worked confidently in the trees within six or ten feet of the observer, but more often they were shy and restless. Time and again the whole flock would mount up from the larches, with a sudden crescendo of their melodious twittering, and circle around high above the trees, before darting down again to resume feeding or departing for another wood. The latter procedure might entail a flight of two or three miles across the strath or over the moors at a height of some hundreds of feet, advertised by their characteristic flight-chatter, which resembled the jingling of sleigh-bells. It was my belief that flocks from several miles around would congregate to roost in one of the groves of larches, for on both spring and autumn evenings I often observed flocks of up to thirty birds mounting to a great height above the woods in which they had been feeding – almost out of sight indeed – and then straighten out on one of their wellknown fly-lines.

By the beginning of March the strength of the wintering flocks would have been reduced by two-thirds, and by five-sixths at the end of the month, when the earliest flower-jewels of spring, the exquisite brushes of the larches, were crimsoning the mossy floors of the groves. Comparatively few pairs of crossbills stayed to nest in Badenoch, but the first fledged young made their presence known vociferously at some date between the end of March and the beginning of May, which indicated clutch-laying as early as the end of February. On 25 April one year two red cocks and three dark-streaked grey-green fledglings visited the garden at Drumguish. Thenceforward the two cocks, accompanied by either one or two young birds, continued to visit the garden daily to feed, inexplicably, in one of the flowerbeds; but no hen put in an appearance until 5 May. The latter's plumage was now so dull that, rather than green, it was a drab olive colour: just as the cocks' red plumage was now also dull and 'rusty'. A week later the cocks ceased singing and came no more to the garden.

It was at this season of mid-May that the larch groves came to life, with blackbirds, song thrushes and mistle thrushes all singing from their lofty perches, and a wealth of chaffinch song on a sunny evening. The squirrel, now shedding its pale fawn winter fur, revealing the rich sandy-brown of its underparts, would also be abroad and watch me for some minutes, motionless, halfway up a larch and aligned to its bole, with an enormous cone in its mouth. Then it would flash into lightning activity, scuttering up to a branch, where it chewed at the broken cone with extraordinary rapidity, as it turned it round and round in its paws, then no less suddenly throw it down and sit forward, with its tail of fine-spun hair stretched along its back and projecting an inch in front of its face, to watch me again. But finally, becoming angry at, or excited by my intrusion, it would begin a staccato stamping with both hind and fore paws, while wagging its tail and indeed its entire hindquarters in its fury and uttering an explosive *choc-choc-choc*.

By the end of May the groves were noisy the day long with the songs and call-notes of coal tits, goldcrests and treecreepers, of siskins and sibillant redpolls, of spotted flycatchers, chaffinches and redstarts, and with the 'stuttering' of fledgling robins. There were days, especially sunny ones, when both groves and pinewoods were never quiet from the high-pitched, linnet-like notes of siskins, ever on the move in the canopy: now perching with shivering wings to break into exultant cascades of twittering song; now circling out slowly in song-flight, with spread wings and tail shining like transparent gold bars against the hot blue sky. It was in the middle of April that a cock siskin had first sung alone from the top of our ash tree (and sang from it every day subsequently), after the initial breaking up of the travelling winter bands of siskins and redpolls. Working industriously, all ways up, at the black alder 'cones' and birch catkins in the glens and on the banks of Tromie and Spey, with black and olive-yellow siskins almost touching the rose-tinted redpolls, the former would maintain ceaseless choral song. This they would do without interrupting their search for food, though occasionally one would perch for a few moments (with wings dropping and sulphurous throat pulsing) to twitter its sweet medley of sprightly, slightly harsh, chattering notes that recalled those of sedge warbler, canary and greenfinch in addition to linnet. Its lighthearted and vivacious stanzas, with their cicada-like concluding trill, were charged with the gay brilliance of life under Mediterranean skies rather than the grey skies of North Britain.

But dominant, both in volume and continuity, to all other bird-notes in the groves was the incessant *vee-tew*, *vee-tew* and reiterated *chüp*, *chüp*, *chüp* of the fledgling crossbills, two or three families of which were being fed by their parents on green cones high up in the larches. Cock and hen, hanging upside down from the feathery green fronds, each fed one, or two, young independently.

In contrast to the crossbills' timetable at Drumguish, it was late May or early June before adults with young began to visit our garden in Newtonmore. Four different families might indeed arrive on the same day. In the middle of June the various families would begin to amalgamate into flocks of up to twenty, though separate families were still to be seen as late as the middle of August. In a fine summer the Newtonmore garden would be noisy the day long with crossbills dropping pine cones on the corrugated roofs of the outhouses, and with the monotonous *vee-tew*, *vee-tew* of their young ones. Hour after hour the latter pestered their parents for food; some typically following the hen, others the cock from branch to branch through the pine trees, for although the four or five or six members of a family would arrive in the plantation together they were likely to split up shortly after arriving, with cock and hen taking so many youngsters each. One cock tended two fledglings for a whole week while his mate and the other three youngsters were elsewhere, and on one evening he was working at the cones on a single pine for two and a half hours. Although the fledglings had been on the wing for some months they were still unable, or unwilling, to forage for themselves. However, it could not be said that the parents paid much attention to their massive young ones which, heavily striped and with feathers fluffed up when soliciting for food, bore a curious resemblance to little owls. Only infrequently — once in ten minutes at most — would one parent alight beside its respective following of young and regurgitate a milky fluid. No doubt the latter did feed themselves from time to time, though it was late June before I noticed that some of them were beginning to wrench off cones for themselves, or pick off a single soft flake from a new cone-spike and swallow it; but even so, an old cone would keep a youngster quiet for only a few seconds before it was dropped in apparent disgust. Thus one young cock, after leaning out from its branch to prise off a cone — and in so doing reveal an apple-yellow rump — spent some time twisting at its stalk and was eventually swinging by one leg with the severed cone in its beak; but then, having flown to a branch with the cone, let it fall and, though peering down, made no attempt to retrieve it.

Nevertheless, the parents' industry was prodigious. For that matter, one rarely saw adult crossbills engaged in any activity other than working at the cones, apart from brief excursions to drink from a gutter or sip raindrops off the leaves of a nearby lime tree. All day during the summer they swung about the pines with their parrot-like acrobatics, nipping off the old cones by their stems. The tightly sealed new cones they ignored, though they would tackle the new larch cones, digging out the seeds from the top of the cone without detaching it from the stem. They were quite fearless, working within six feet of the observer with tremendous energy and at great speed, hurling down their cones almost as soon as they had nipped them off. A single cock, feeding only one young one, might drop twenty-eight cones in twenty minutes, and within a few minutes of the first family's arrival in our plantation the lawn was littered with a couple of hundred cones, while in the course of their first evening the six parents of three families dropped more than five hundred cones from the twin pines that towered over the roof of the house.

The larch groves were still full of sound in July, with the 'reeling' and *chuvee*ing of coal tits swinging through the trees; thin whistles of treecreepers working up the boles in twos; sweet whispers of willow warblers; an occasional twitter from a passing siskin or redpoll feeding young; the emphatic 'stuttering' of flycatchers also feeding their piping young, and the no less emphatic stuttering of parent redstarts. But with the departure of the crossbills, usually late in July or by the middle of August, the groves declined into silence, for the siskins were quiet by that time, and one heard only the small whisperings and trillings, twitterings and 'saw-sharpenings' of coal and crested tits, goldcrests, redpolls and willow warblers. If a mild winter followed and no crossbills returned to take up winter residence, then the groves would be silent for another six or seven months.

8 Inhabitants of the Planted Woods

ONE OF MY ambitions as a naturalist, before we went to live in Badenoch, was to explore what remained of the old pine forests of Caledon and find out precisely what birds and beasts inhabited them. It soon became evident that the ecology of the planted pinewoods around Drumguish was not typical of the native forest. Capercaillie rarely came near them. During the non-breeding season one might flush an occasional brace of blackcock from the old heather in boggier places, where the pines were bigger and more widely spaced, but the greyhens never visited them. Coal and crested tits might be encountered in every month of the year, but I found none nesting. Crossbills only alighted on the canopy when flighting from one grove of larches to another.

The reasons for this shunning of the planted woods were patent. The planted pines grew so close to one another that any considerable lateral branching was prohibited. Even at midsummer only palely diffused rays of sun penetrated their twilit interior. One could walk through a hundred acres of the latter and meet with only a couple of small bands of tits and goldcrests foraging high up in the pines. No vertebrate creatures (and few invertebrates) lived in the heart of the woods. No flowers grew there, though feathery mosses covered every inch of ground except trodden paths. There were few days in the hardest winter that snow filtered through the canopy heavily enough to shroud that warm evergreen carpet of moss. Lichens climbed six or twelve feet up the boles of the pines, and in autumn the beautiful orange stag's-horn fungus sprang from decaying logs. The pinewoods, and indeed much of Badenoch from the floor of the strath to over 3,000 feet, supported an incomparable wealth of fungi of every imaginable form and colour — scarlet, brown, white, lilac, green, pale purple, orange, red and orange, black, yellow or tangerine.

In plantations that had not been thinned there was barely room to sidle between one tree and its fellow, and all sense of direction was lost only a few yards in from the edge of the wood. These were woods of death, strewn all ways with hundreds of fallen trees, while those standing were suffocated by a monstrous greyish-white lichen and grotesquely festooned with lacy strands and cobwebs of another grey-green lichen, which hung in leprous strips from every branch and snag: a cancer of darkness — the Spanish moss of a Nicaraguan jungle or of a Louisiana *chenière*. In them there was absolute silence. No birds sang, no insects hummed. Alone of all flowering plants the wood-sorrel carpeted the dead brown floor on extreme southern edges, where yellow sunbeams sometimes slanted in. The wood-sorrel: that delicate little plant that covered scores of square yards of shaded ground beneath the giant spruce firs, more than one hundred feet in height, on the gloomy western fringe of the Drumguish woods, where not even mosses grew and where in autumn the few sun-starved fronds of bracken were the palest of browns, greens and straw-yellows. Each leaf of the sorrel was folded down in three heart-shaped sections, thus forming the bit of a half-inch brace, which opened when the westering sun streamed into the wood in the evening; but only here and there did a fragile white star-flower appear.

No flowering plant sprang from the soft, deep green carpet of mosses, littered with pine twigs and branches and drifted with tawny layers of pine-needles. It was only on the verges of the woods, or where pathways and little glades permitted a fuller measure of daylight to penetrate the canopy, that a vigorous growth of cowberry and some blaeberry flourished, but rarely flowered and fruited; and where the berry-plants grew, there the few flowers of the pinewoods bloomed. First, in April (before the wood-sorrel) the ethereal mauve-flushed stars of the wood anemone, which, however, loved better sunny banks and damp birch groves; then, early in June, solitary plants of the third and most perfect of this beautiful white-star sequence of the pinewoods, *Trientale*, the chickweed-wintergreen, whose six-pointed flower-head rose on a thread-like stalk two inches above a whorled crown of long, slender, slightly drooping leaves. Its tall yellow anthers contrasted with its delicate white petals, which in many plants were exquisitely veined and tipped with a lilac-pink tint — as were in some cases the leaves and stem. It, too, preferred the birch groves. In July *Trientale* was succeeded by the true (intermediate) wintergreen, whose white and rose-tinged inverted globes hung from a six- or nine-inch stalk; and here and there the cow-wheat displayed its pale yellow trumpets. Finally, at the end of July, the last of the pinewood flowers, the white orchid-like lady's-tresses, thrust up its twisted plat from paths strewn with brown pine-needles.

Just as the flowers grew only in glades and at the fringes of the woods, so insects were to be found where there was at least a modicum of light and sun. Spiders slung their gummy webs from silken cables swung between two trunks at a height of five feet above a path. The queen *Bombus lucorum*, returning to a pinewood ride from her first April expedition abroad, alighted at the entrance to her chamber in a mossy bank at the base of a pine. But the dominant invertebrate of the pinewoods was the wood ant. Lined up along a few score yards of fencing, or at the bases of pines, at the edge of a glade and especially a path, might be as many as five or six rounded, cone-shaped mounds of pine-needles, varying in size from a small cushion to monsters twenty-seven feet in circumference at the base and four feet in depth. These orange-brown 'hills' were the most notable feature of the pinewoods. Throughout the winter the black and brown inmates remained under cover, though there were always a few sluggish ones to be found in the soft dry mould within a quarter of an inch of the outer covering of pine-needles; but on a mild humid day in March the tops of those hills free of snow swarmed with ants, and a few might be observed moving slowly into the undergrowth. Even at that early season a jagged mound was busily repaired, with the inhabitants carrying needles, cone-flakes and other debris to the centre of the flattened hill. Needles up to an inch in length were carried in their jaws, while larger 'twigs' were pushed or pulled by an ant at either end, assisted perhaps by one in the middle; but there was no apparent work-plan among the seething mass. A creditable out-of-season job was made of these repairs, though the final product was ragged and layered, and lacked the smooth finish of the perfect hill.

On an April day of blue skies, when broom pods were popping and dead pine-cones crackling in the hot sun, and the less densely crowded younger plantations were full of the soft twittering and merry jingles of travelling siskins, pricked out in gold, the dome of a sunny anthill would be seething with a black mass of ants 'inches' deep, though the inhabitants of shaded mid-wood hills would still be sluggish and mostly interred; but by May all the ants would be venturing far afield and adding the new season's quota of pine-needles to the hill. At mid-June, when the pines were redolent with the hot, spicy, masculine scent that breathed the sun-dried, tindery aroma of summer, endless patrols marched along the 'roads' of springy brown peat swept as bare as a beaten earth-floor. It was not clear whether those ants constantly 'sweeping' the roads, immediately pulling away any pine-needle that fell on the thoroughfare, were road-sweepers by profession or any ants that chanced to be passing at the time. Large twigs were circuited or undermined, as were experimental obstacles in the form of big stones or pits, which caused little confusion or delay to the interminable processions. The busiest roads, scores of yards long, more than a hundred in some instances, radiated from the swarming hills to the tallest pines on which no lichens grew. Up and down the rough-barked trunks — up and up to a height at which my binoculars could no longer follow them — passed in continual succession two streams of ants. Those descending did not carry any obvious booty, since the majority of them had no doubt been 'milking' aphids of their 'honeydew'.

As in the case of flowers and insects, so with birds. It was along the fringes of the woods and in inhabited clearings that the few blackbirds, robins, chaffinches and tits were to be found, while on any day from February to May one could lean back against a pine tree and listen to mistle thrushes throwing their strong, clear whistles against the soundboard of the encircling woods. As they answered one another, close above my head, I learned that the fine 'remote' notes, which at any distance appeared to comprise the full song, were in fact linked by a rather harsh twittering. So sweet and throaty were the notes of these Highland mistle thrushes, and so thin those of the blackbirds, that despite the characteristic phrasing common to all British mistle thrushes, there were occasions on which I declared an unseen thrush to be a blackbird. They were incomparably superior singers, and almost as contralto as an English blackbird.

It was the end of March before the first of the resident blackbirds would make a halfhearted attempt to tune up, for of no songbird was it truer that intensity of song depended upon concentration of numbers. There were several blackbirds in the vicinity of Drumguish, but they were scattered over half a square mile and never sang with anything approaching the continuity or flute-like tone of their fellows in the garden suburbs high on the hill above Kingussie. Even in May a Drumguish blackbird would perch on a lateral pine bough on the extreme edge of the wood and listen all day to the mistle thrushes, without ever uttering a note. Only in the latter half of that month, when the young thrushes were fully fledged, did their parents at last surrender song to the blackbirds.

Although tawny owls inhabited a dark avenue of spruces, their kind had never apparently been commonly distributed in Badenoch. The keepers' guns were against those that did live in the strath, and the very few that survived were shy and, lacking the stimulus of numbers, rarely, very rarely, enhanced with their hooting moonlight nights in the pinewoods. Theirs was a night sound that I missed as much (remembering their joyful yodelling from the wooded banks of Loch Awe and the wooded glens of Kirkcudbright) as the songs of warblers and thrushes by day.

Two or three red squirrels also frequented the avenue of spruces, attracted by food put out for them on a cottage windowledge, but they were not numerous in Badenoch, and were seldom to be seen in the old pine forest in the high glens. There were some years when I did not see any in their usual haunts for periods of weeks or months at a time; then, one summer, there would be an 'invasion' on a very small scale, and a few might remain throughout the following winter. As in other parts of Britain the reasons for the fluctuations in their numbers were obscure, though in Badenoch the issue was not confused by the presence of grey squirrels, which had never colonised this part of the Highlands. Nowhere else in Britain had squirrels a more dependable food supply than in Badenoch, which provided a perennial and unfailing store of various cones and lichens and, in their season, an exceptional abundance of fungi. The latter were held in the paws, with one end resting on a branch, while the squirrel nibbled away with characteristic urgency; only to leave the agaric lying on the branch and streak up to the top of the tree on hearing the guttural alarm-note of another squirrel. Nor did the severe Badenoch winters appear to affect the squirrels. They were, for example, out and about in temperatures as low as minus 10° F to minus 15° F for at least the first forty days of the phenomenal freeze-up in the New Year of 1947, when there was partial snow-cover over all Badenoch, and when many of the pines were singed, broom and juniper killed off, and thousands of acres of heather frosted; and their numbers were at a normal level the following autumn.

Too often I found squirrels that had been hit by cars, but I never found one that had died a natural death or from disease — one would hardly expect to. With virtually no predatorial enemies at the present time in Badenoch they ought to have multiplied; but if their numbers did not often fall below the norm, they on the other hand never rose above it except during temporary invasions. That being so, the factor limiting their population had to be connected with their breeding cycle or with the incidence of epidemics such as cocidiosis, as had probably been the case in other parts of Britain since the turn of the century.

In a mild winter the period from autumn on into the New Year was the squirrels' season of maximum activity, and on an October morning there might be as many as five in the crossbills' grove of mixed larch and pine at Drumguish. They were no doubt a family group, and much fun and games they had, chasing through the trees in twos and threes with excited squeaks and muffled gutturals; each on the other's tail, round and down a trunk, and up and round with a brittle scrabbling of claws on the tindery bark. They would spiral round the trunk, as if swarming up a greasy pole, with such rapidity as to put one in mind of whirling golden-red catherine-wheels. (Why were so many British mammals a fiery, foxy, ruddy brown in colour?) Then on an instant, with startling abruptness, one squirrel would freeze in a pose of luxurious abandon, hanging head down from widely spaced and fully-stretched hind legs, with front paws dangling and bushy tail aligned to the trunk. How characteristic of squirrels was this sudden transition from hyperactivity to frozen inertia in the most improbable position and even in a strong wind, and despite the fact that the animal might be holding a huge cone in its mouth.

The graceful agility of a squirrel among the slenderest twigs was past perfection: an effortless removal from one point to another with such deceptive speed that the eye followed its movements with difficulty, as it swarmed up the smooth bole of a tree without the slightest break in the rhythm of its ascent; then out along the swaying twigs, and a flying leap to the equally slender fronds, or even to the trunk, of a larch on the other side of the ride. (Did squirrels ever fall?) There it would peer out for a cone, twisting itself half into space and, having torn one off, retreat to a safer perch. Presently its mate would appear and a shower of cone-flakes would be hurled down on me. Then one would scuttle down with a cone and skip off to bury it; but having scraped a hole, fill it with leaves and place the cone on top of them, before returning to a nearby log and watching me with its protruberant round black eyes, while scratching its head and curving its tail over its head.

9 Birds of the Old Pine Forest

DRUMGUISH was not very far from a remnant of the old forest of Caledon. From our windows the tops of the highest pines in Badan Dhu in the lower reaches of Glen Feshie could just be made out as a frieze along the crest of the hill less than two miles over the moor as the raven flew, though very much farther as the naturalist went.

Badan Dhu was a jungle of waist-high heather, dense though spindly blaeberry shooting up to a height of two feet, brushwood and twisted limbs of felled pines, shrubby ten-foot junipers and pockets of young self-sown pines. When toiling up a forestry track through this in places impenetrable tangle I would halt from time to time to listen and, as often as not, there would be the sudden muffled crash of a heavy form breaking away from the undergrowth behind the screening trees; but I would see nothing. For a bird the size of a turkey the capercaillie's elusiveness was remarkable. Indeed, so long as one kept moving, a caper preferred to stand motionless on its pine tree or sit tight on its pine-needle 'form' at the base of a tree. The latter was a favourite resting place, and the rough-textured droppings of pine-needles were to be found everywhere in such sites. So close would a caper sit that one might actually tread upon the bird before it exploded, as I did one June day when struggling across country to Loch an Eilein through miles of felled woods and old shrub-like heather, and a cock burst so violently from the undergrowth that he left a big blackish-grey feather beneath my foot.

Stand and stare at nothing in particular and it was a fifty-fifty chance, if I had not allowed my concentration to wander, that I would be rewarded by a glimpse of a cock caper with a metallic dark green patch on his breast, a grey back and an enormous black tail, vacating his low perch in a nearby pine; or perhaps it would be a hen creeping out from under a log, like some huge hen pheasant, for her plumage was a dull old-gold with white speckling on the back, a tawny-chestnut patch on the throat, and a copper-edged tail. If I was especially fortunate the cock would alight on the topmost branch of another pine and stand there, watchful, for minutes. Cormorant-like of swollen head and neck, he was predatory of aspect, and as dark as the pine against a grey sky, except for the white fleck at his shoulder. Treading cautiously closer from tree to tree I might eventually get near enough to make out the vivid small points of red at his eyes and his bright yellow beak. The latter was a splash of colour visible at a great distance on a sunny day, when one zoomed high over the canopy of young pines in a mile-long glide below the crags of Badan Dhu.

Above all else the caper loved a bushy heather undergrowth, and it was in the heather park of young pines on the northern edge of Badan Dhu that I could always be sure of flushing one. In March and April I might put up as many as nine solitary capers, and in June as many as eleven hens (six in a single pack); but from June to August no cocks, or blackcock for that matter, were to be found in Badan Dhu. Whether all deserted the forest at that season or merely went into close hiding during the moult I was never able to determine, though some certainly left the woods temporarily. It was a sign of the seasons when one or two blackcock made their appearance at the edge of the Drumguish woods in July, and hen capers with young might sometimes be seen feeding on the moors where heather had been burnt in previous years. The cocks were more difficult to locate, but one did occasionally flush them from unexpected places. Thus, one mid-July afternoon I was sitting high on the steep slopes above Glen Tromie, observing at one and the same time a blue hare — a flat gritty summit being a favourite resort — a roebuck bounding down the hill, and a bright chestnut fox cub, which was watching me attentively, with ears cocked. A cock ring-ousel was 'tchacking' from a nearby rock; a blue peregrine, circling about the cairns below me, 'whickered' incessantly; a kestrel mewed from the hillside below the peregrine's eyrie. Suddenly, to my surprise, first one and then another cock caper rose from the hillside and zoomed out over the shimmering silver ribbon of the Tromie, 700 feet below, to the birchwoods on the other side of the glen. From my high seat I seemed to be gazing down on a dream forest rather than one of reality, for the dark green canopy of birches was hazy in the bright sun. Two herring gulls, migrating up the glen, let fall the sudden laughter of the sea as they passed lazily south.

In addition to the few capers, blackgame and crossbills there might be individuals of four other species — wren, bullfinch and coal and crested tits — in Badan Dhu on a winter's day. All four had inhabited the old forest of Caledon. Wrens were still to be found in those parts that had not been felled, mainly in the lower reaches of the forest in the vicinity of burns, though in those that had been heavily cleared their habitat extended right through the forest from top to bottom. Their distribution in Badenoch was curious. I knew of no more than a score of pairs on my Drumguish beat, only one of which nested anywhere near man's habitation. Moreover, although these breeding pairs ranged from 750 feet in the birch parks fringing the Spey meadows, and up through the glens to 2,000 feet in the remotest *allts* of the Forest of Gaick, the central massif of moorland between the glens of Tromie and Feshie, with its attractive tree-lined burns and little glens, was destitute of them. During the period October to January, however, they visited the Drumguish moor, and then for some weeks in January and February all would disappear from my beat. Naturally, since their numbers were so sparse, in contrast to every other region known to me in the British Isles, their movements were secretive and their songs seldom heard, not sustained and thin in tone.

The gorgeous bullfinch was also an enigma and a persistent wanderer. Indeed, the life history of the Highland bullfinch was one of the least known. It was a little after dawn one November morning, when the savage storm of the previous twenty-four hours had abated, that I heard a soft piping among the birches, and there to greet me was a troop of five bullfinches, four of them brick-red cocks. They seemed surprised to see me, tilting their black-capped heads to look up at me, while drinking from a pool at my feet. I stood quietly watching them for a short while, fascinated — as always — by their harmonious shades of black, pink, white and grey. I could not linger with them, but I thought about them all day, and for several days afterwards: not only because of the impact of their beauty on a particularly grim Highland morning, but also because one did not encounter bullfinches very often in Badenoch. They nested in few places outside the pine forests, and when they did so it was predominantly in spruce plantings. Throughout the year, no matter how severe the winter, their headquarters were the native pine forests in the high glens of Feshie, Rothiemurchus, Glen More and Abernethy. Carpeted with rough heather and blaeberry between the great spreading trees, an old pine forest felt warm and dry in the winter; and an otherwise unrewarding day for a naturalist was always made memorable by even a fleeting glimpse of a bullfinch, with its rose-coloured breast and flat velvet-black cap and short black bill. How pleasant it was, when skiing over the moors through a maze of Christmas trees, to hear all around me the soft piping of bullfinches trooping from one outlying pine to the next through the drifting snowflakes. Such winter flocks commonly included six or seven individuals, though occasionally as many as fifteen or thirty, in which cocks usually outnumbered hens.

It was in September that small parties of bullfinches dispersed from the forests into gardens and along the burnside alders, to forage among the birch scrub and bushy heather on the moors. It was typical of these Highland bullfinches that when one was out watching deer, with only grouse for company, one should chance upon a flock migrating from one pinewood to the next through a treeless pass at an altitude of 1,700 feet. Some of these wandering flocks were no doubt composed of true migrants from northern Europe, for in October there were always bullfinches among the Scandinavian thrushes migrating south into Perthshire through the storm-bowed rowans and birches in the Forest of Gaick, which was several miles from the nearest bullfinch country; and on the road to Gaick I watched on one occasion what were almost certainly two splendid cocks of the northern race feeding on heather seeds protruding from a drift of snow. In short, the habits of these Highland bullfinches were similar to those of their relatives in northern Europe, and they were therefore to be regarded as natives of the Forest of Caledon, rather than as descendants of immigrants from southern Scotland.

The coal and crested tits were the most constant inhabitants of the native pine forest, though even long-tailed tits foraged in autumn as far as the ultimate outlying pines in such high glens as Gaick and the Lairig Ghru. No doubt the deeply corrugated bark of the old pines harboured abundant supplies of insects and their larvae for tits, goldcrests and treecreepers, exploring upwards to the tree limit between 1,500 and 2,000 feet. Nevertheless, one could not but marvel that such delicate mechanisms could be adequately fuelled at midwinter when temperatures were as low as minus 15° F.

Rather few pairs of long-tailed tits summered in Badenoch, though towards the end of May one might chance upon a number of them in desperate combat, not only falling through the air while grappling with their claws but continuing to fight on the ground. However, from September to December in some autumns extraordinary numbers of them roamed through the remoter wooded glens and passes. These roving bands of twenty or thirty long-tails were therefore probably immigrants, possibly from Scandinavia, though they did not include any with wholly white heads typical of the north European race. I passed hours watching them in the Craig Dhu birchwoods and listening to their sibillant, trilling choruses of alarm when a buzzard hovered low over the trees or when a sparrowhawk rocketed among them. But though they streamed south-west in scores from tree to tree, I was never able to decide whether this was a true migratory movement or merely a localised one in search of food; for once they reached the southern edge of the woods and were confronted by a mile-wide expanse of bare hill and bog, numbers of them would turn back north-eastwards again, and I observed this same two-way traffic through my garden in the village.

As early as the third day of setting up home in Drumguish I heard a soft trilling song in the pinewoods that was new to me. Four days later I identified the singer, which was hammering away at a grub under its foot on a pine branch. It was a plump, squarely-built tit with a short forked tail and a brown back, yellowish underparts, an intensely black, wedge-shaped gorget and a unique shovel-shaped, bluish-grey or purplish crest. The purring trill — a soft 'water-bubbling' not unlike a greenfinch's 'chirrup' — was not, however, the crested tit's song, but its contact-note, which was to be heard from almost every mixed band of tits throughout the year. Actually a full twelve months had passed before on a day late in February I chanced upon a party of half a dozen coal and crested tits in some outlying pines on the moors and listened for the first time to a proper song from the latter, as the band flitted from branch to branch and tree to tree on their ceaseless search for food, picking out half-inch grubs from the deep corrugations in the bark. A sweet and throaty phrase of seven or ten notes, which could be represented by the syllables *blae-berry*, it was as pure and musical a dunnock-like song (but with a richer warbler-trill) as I could remember hearing, and quite dissimilar from the purring trill which the singer subsequently uttered persistently. The accompanying coal tits were also in a state of intense excitement, expressed by a prolonged 'reeling' similar to that of goldcrests though with a grasshopper warbler's electric emphasis. I concluded, therefore, that the presence of my dogs was the cause of this excitement, and that this infrequently heard song of the crested tit was incited only by some unusual external stimulus.

My second experience of this particular song — neither confirming nor contradicting the above conclusion — occurred in August, when a travelling band of more than a hundred tits, mainly coal and crested, were in the Caigeann, and one of the crested tits momentarily interrupted its foraging to perch on the branch of a pine and utter a phrase of song.

My third experience was early one January, when a thaw had broken a hard black frost. In this instance some scores of coal and crested tits were darting in continual succession — characteristically — from pine to pine with excited calling, which included two brief bursts of song from a crested tit. Since the coal tits were also calling to one another with their clear, nostalgic *chuvee, chuvee, chuvee* for the first time in the New Year, I assumed that the fresh weather, and not my presence, was responsible for the crested tit's outbursts.

My fourth and last experience over a period of two years was again in February when, after a light fall of snow in the morning, I was watching a blackcock *rookoo*ing from the top of a dead pine on a mild afternoon, and heard four notes of the sweet and brilliant warble.

It was evident, then, that this uncommon song of the crested tit was not a true song associated with the breeding season, but a response to such stimuli as a change in the weather or the appearance of a strange object, and could perhaps be compared to that primitive, prolonged, lisping song of longtailed tits when alarmed by the passing of a sparrowhawk along the edge of the woods. The crested tit did, however, have a true song, resembling the blue tit's bell-song. This could form the prelude to the purring trill, and might be rendered *see-see-see-see-whorr*. It was to be heard very frequently during the breeding season up to the middle of May, and occasionally during the winter — as, for example, from an individual in a travelling flock in mid-February (when these tits were first in pairs) at the end of a week's thaw after January frosts. However, it could also be provoked by external stimuli. In one instance the cock of a pair, building in a cavity eight feet up the rotten skeleton of a pine, was excited by my presence near the nest-hole. Flitting agitatedly from tree to tree (under one of which a cock capercaillie was sitting), or perching on a branch, it uttered incessantly its squeaky alarm trill; and this intermittently crescended into the rattling *see-see-see-see-whorr*.

There were three pairs of crested tits collecting building material in the Badan Dhu pinewood that May, vigorously flaking bark off rotten boughs; but time and again the hens, though only occasionally soliciting the cocks, would shiver their wings, while uttering a faint, tremulous trilling, pitched as high as a goldcrest's 'reeling'. At the beginning of June, when only the purring trill was to be heard, the tits were feeding nestlings; and before the middle of the month families of both crested and coal tits were already wandering through the dark pine forests, and the crested juveniles, with their short, ragged black crests, would perch inquisitively within four feet of my face, purring metallically.

Of all the old forest birds the crested tits were the most attached to the pinewoods. However they, like their constant companions the coal tits, might be found in a spur of self-seeded young pines extending out into the moor from the main tract of forest every day right through the winter, and they would often flit down from the trees to pick up a seed or invertebrate in the heather. Moreover a crested tit, accompanying a roving band of all the other tits, might even fly down from the edge of a pinewood to hack out a grain of oats from the inverted sheaf crowning a stook in a cornfield on the other side of the road, and fly up to the trees again with its single seed. But I never found them in the larch groves, though they would examine the young green cones of a larch in a pinewood.

In Badenoch we were on the extreme southern edge of the crested tit's Highland range, and though I encountered them from time to time in the birches below the pinewoods or travelling along alder-lined burns through the moors from one pinewood to another, they did not often rove outside their regular habitat. Nevertheless, they were perhaps extending the bounds of this, for I found colonies in Glen Shirra at the approaches to the Corrieyairack and in the extensive plantings of great spruces at Ardverikie above Loch Laggan. These colonies were twenty miles from their nearest haunts in Glen Feshie.

10 Bees on the Moors

ON SOME morning between the middle of March and the beginning of May, depending upon the vagaries of a particular spring, the first queen small-earth bumble-bee (*Bombus lucorum*) might be observed combing herself in a little hollow in the moss at the base of a pine tree, not omitting any part of her black body with its burnished yellow or golden-brown bands and white tail. The earliest of the wild bees to awake from five or six months' hibernation in a mossy hole, *B.lucorum* was also the most widely distributed of the three most numerous species of bumble-bees in Badenoch. The other two were the bilberry bee (*Bombus lapponicus*) and the moss-carder (*Bombus muscorum*), though numbers of the gypsy cuckoo-bees (*Psithyrus bohemicus*), which were parasitic on *B.lucorum*, were to be seen in flower-gardens, often in company with their ill-fated hosts, on fine days from May to late August.

Once the *B.lucorum* queens had emerged from their retreats — very possibly after the heaviest snowfall of the spring! — they were abroad in all weathers, visiting the bearberry's bulbous rose-pink and white flowers, which clustered in groups of three to five among the mat of shiny, dark green leaves. On a day of strong cold winds in the first week of April they might be on the wing as high as 3,000 feet, and by the end of that month be zooming over the 4,000-foot summits.

Complementary to the healthiest and most vigorous growth of heather I had seen anywhere in Britain was that astounding carpet of berry-plants, which together with an abundance of lycopods, scarlet, black and silver lichens, and the false reindeer-moss, thrived on moors that were predominantly dry despite a few peat-hags and sphagnum-bogs. The berry-plants flourished especially along the verges of moorland roads and on flats and protruberant knolls, actively colonising those areas that had been temporarily denuded of heather by muir-burning, though never stifling the new growth of that extraordinary plant. A perfect symbiosis existed between heather and berry-plant — members of the same family — and between berry-plants and lichens. When the low December sun, only just clearing the rounded top of a hill to the south-west of the Drumguish moor, lit up the scarlet berries of bearberry and cowberry in vivid little motes of colour, whole patches of moor, particularly hollows in the heather, blazed with the red 'sealing-wax' tips of the *cladonia* lichens, small colonies of which shot up from the mat of bearberry and cowberry and the ash-green filigree-work of the reindeer-moss, another *cladonia*. Although the latter might be the predominant vegetation on a plateau of granitic gravel at 2,000 feet, it was also plentiful on grassy and gravelly braes down in the strath. So too, the scarlet sporophores of the *cladonias* were to be seen as high as 2,800 feet in any month of the year.

The commonest berry-plant was the bearberry (the *uva-ursi* of the Canadian barrens, whose flora so much resembled that of the Grampians). In Badenoch both the bearberry and the almost as extensively distributed cowberry were known as the cranberry, which was, however, scarce outside Rothiemurchus. Inextricably mixed with the bearberry, with all four plants thriving in the same tough creeping mat of that plant, were no less vigorous growths of cowberry, crowberry and in some places blaeberry (the whortleberry or bilberry). The bigger and paler green leaf-shoots of the cowberry stood up from the dark green mat of the bearberry, and contrasted with the crowberry's spiky yellow-green leaf-whorls, which almost concealed the minute, dark purple flowers in their axils. Spiders lurked in silken chambers opening on to webs stretched across the clumps of bearberry.

In May the *B.lucorum* and *B.lapponicus* queens celebrated the golden birth of the sallow. Normally easy-going buzzers from flower to flower, there was now an almost vicious frenzy in their hum and an intense excitement and urgency among all present, comparable to that of hive-bees swarming. Also at the sallow spindles would be one or two black dung-flies with orange wings and numbers of a solitary bee – the early mining bee which, though resembling a small bumble-bee, was instantly recognisable by its fly-shaped face. It was the second or third week in May before the first hatch of diminutive *B.lucorum* workers were to be seen at the feast of the sallow; but that was only on sunny days, when they were on the wing much earlier in the morning than the queens, which appeared to undertake most of their foraging on dull days.

By early June the florescence of the sallow was nearly over, and the prevailing berry-plant flowers were now the white trumpets of the cowberry. Although the yellow kidney-shaped flowers of the aptly-named needle-whin — widely distributed over the moors up to 2,500 feet — might appear as early as the third week in April the general flowering of the very special moor and bog plants coincided with the emergence of the worker bees. Earliest, in a boggy place, a pink lousewort or perhaps a colony of white ones; then the cherry-pink tuberous pea, which was peculiar in fading to blue as it quickly wilted, and an unexpectedly delicate plant to find growing in a few places among the strong, coarse foliage of bearberry and cowberry at the edges of moors and as high as 1,750 feet in the sunless pine forest. The mountain-everlasting, with its white or more beautiful rose-red flower-buds, was to be found, like the needle-whin and milkwort, at over 2,000 feet and, as solitary specimens, at 3,000 feet. One had at first to look closely for the two long, winged, purple-blue sepals and lilac-pink corolla of the milkwort, which was soon to bejewel every square yard or moor, where the heather was not too thick. With its wide range of colouring — a feature of the moorland flowers — deep purple-blue, purple-blue with pink corolla, light-blue with white corolla, pink, pink with pale-pink corolla, or pure white, the milkwort achieved perfection with a royal Tyrrhenian purple of the deepest hue; and as plants of this colour thrived more vigorously than those of other colours, clusters of these royal purple flowers formed vivid patches on the short grass and heather of the braes.

The exquisite purple of the butterwort, which grew in colonies on the moors and to 3,000 feet on the tops, was a little paler in hue than the milkwort, and the plant owed its supremacy in beauty to the soft velvety texture of the perfect little flower crowning the long stalk. Purple was indeed the dominant shade of the dwarf moorland flowers — the yellow tormentil excepted — until June, when the lemon-yellow flowers of the rock-rose, with their glistening orange centres, bloomed everywhere on green knolls and boulder-studded braes, and the white cruciforms of the honey-scented heath-bedstraw laced greensward and moor in extraordinary profusion to over 3,700 feet on the tops. June was also the flowering season of the broom. The crumbling stone-dykes, with their luxuriant yellow-green clumps of male- and lady-ferns, were festooned with the broom's globes of lemon-yellow flowers, murmurous the day long with hive-bees; and when I came down from the hills in the grey light of dusk the township's lanes would be glowing bronze-yellow.

By the middle of July the moors, up to 3,700 feet, were carpeted with eyebright (simulating white heather) and edged with banks of lilac thyme and flame-coloured and yellow St John's-wort. There would be a pink cast of cross-leaved heather on boggy flats and a purple spread of bell-heather on gravelly braes. On grassy slopes the field-gentians thrust up their pale violet poker-heads from dull green sheaths. Meadows and hedgebanks were gay with masses of lady's-bedstraw, sowthistles, spearworts and buttercups, harebells, violet tufted-vetch and the purple heads of the giant melancholy-thistle.

It had been stated that once the worker bumble-bees were on the wing they gradually took over foraging duties from the queens until ultimately the latter no longer ventured afield. But such a life history was not compatible with my observations on *B.lucorum* in Badenoch, where, as a general rule, queens were in the vast majority until the end of June. Moreover, in some years very few workers of any species of bumble-bee were to be seen anywhere in Badenoch until September. Although ranging over the highest mountains, *B.lucorum* was essentially a bee of gardens and heather moors. When traversing through the Forest of Gaick into Perthshire one passed through zones of grass and heather alternating with scree slopes of blaeberry and wild raspberry. On a September day the screes were the foraging grounds of *B.lapponicus*, but as soon as one entered the heather zone again the predominant bee was *B.lucorum*. Workers of both species, together with a few queens, would be working feverishly in those last days of summer, though in a fine autumn large numbers of *B.lucorum*, both workers and queens, might still be visiting the heather 2,000 feet up on the high moors as late as the middle of October; but by that date all were sluggish, constantly settling to rest in the heather, and by the closing days of the month the last young queen would have crawled into her winter retreat.

The life history of *B.lapponicus* conformed to the stated pattern. The earliest amber-tailed queen might emerge from her burrow under a stone in a tough mat of bearberry by the end of the first week in April — though it was often mid-May before she did so — and by the end of May single queens might be continually passing over the 3,000-foot mosses of the Cairngorms on a swift and direct north-east to south-west fly-line or on this bearing in reverse. What was the purpose of these expeditions, which must have carried them unusual distances from their nests? In July even the tiny workers (first on the wing in the third week of April) visited the clumps of moss-campion in the 4,000-foot corries. Although the flowering-currant attracted the queens — as it did all bees — they did not linger long at the garden flowers, but at the bell-flowers of the bearberry on the moors and at the golden spindles of the sallow they might outnumber the

B.lucorum queens by ten to one. My notes indicated that few *B.lapponicus* queens were to be seen after the end of May, from which time until the first week in September the workers were responsible for all foraging.

The striking mustard-yellow moss-carder queens, with their distinctive, dark chestnut thorax saddles, were the latest to emerge from hibernation in the middle of April; the workers, appearing a month later, were on the wing until the first week in October. This was a bee of the gardens and lower moors — I never observed one above 2,000 feet — and was the only insect, apart from *B.lucorum*, that visited the honeysuckle in our garden.

Every spring I found myself speculating about the origins and ultimate destinations of those small tortoiseshell butterflies that were to be seen at 3,000 feet on the high tops as early as the first week in April, only a fortnight after their initial appearance in the strath gardens. In some years there were quite remarkable numbers of them far up into the hills by the middle of April, while in July and August they were to be found as high as they could be found in Britain, visiting the moss-campion above 4,000 feet. On one August afternoon I counted more than a dozen flying in all directions over the Wells of Dee plateau below the summit of Braeriach, though, as in the case of the bees, north-east to south-west was the predominant flyline; and they were still to be seen on the high moors as late as the first week in October.

One was also as likely to see painted ladies 2,000 feet up on the moors as in the strath; but they were rare migrants to Badenoch, and I recorded them in only three years — between the last week in May and the middle of September. Nor did many red admirals visit us. Indeed, although we might expect to see faded specimens from the last week in May until the end of June or the middle of July, and subsequently newly hatched ones from the first week in August until the end of September or exceptionally the middle of October, I had almost forgotten during our years in Badenoch that a butterfly of such gorgeous beauty actually existed. Thus when, after sixteen years, as many as ten red admirals — all perfect insects — were to be seen on a single dahlia plant in the garden early in September, and later sixteen on a cluster of Michaelmas daisies, I was able to indulge my primitive delight in bright colours to the full. As it happened, none had been seen that year prior to September. Where had they been hatched?

It was also in September one year, when we were living in

Drumguish, that we were visited on the first of the month by a hummingbird hawkmoth — and also by the first red admirals and painted ladies of the year. While hovering on ceaselessly vibrating wings the moth would drive its long curved 'watch-spring' tongue deep into one phlox flower after another, the cherry colour of which matched its underparts. When it finally wheeled away from the phloxes with lightning acceleration it resembled a true hummingbird, with curved tongue (for beak) as long as its body. It visited the garden again on each of the next two days, and in the cold grey stormy evening of the fourteenth entered the house and stayed the night. Thereafter it disappeared. Imagine then my astonishment on 10 May the following spring when, with the temperature well up in the sixties on a third summer-like day, there was once again a hummingbird hawkmoth in the garden at noon, imbibing deeply from the polyanthuses. Would anyone gainsay that this was our over-wintering moth?

Some years later, on 2 June, when it was sunny, though cool and windy after a fine May, what appeared at a casual glance to be a large, pale yellow bumble-bee was hovering over the aubrietia; but since such a bee would have been most unusual in the Highlands I examined it more closely. The blackish margins to its quivering wings and its black, horned antennae testified that it was the narrow-bordered race of the bee-winged hawkmoth.

11 Dragonflies in the Pinewood Peat-bog

IT WAS on the hot summery morning of 10 May one year, when lizards were sniggling into the heather from pinewood paths, that the first olive-brown, four-spotted libellula dragonfly (*Libellula quadri-maculata*) of the season rose from the bushy heather in a peat-moss on the sunny south flank of the Milton pinewood, a few hundred yards along the edge of the moor from Drumguish. Immensely broad-bodied, its wing formation curiously reversed the familar delta-wing of a jet plane; but its powers of flight were no less efficient because of that, for this superb aerobat could, like a hummingbird, reverse direction in flight with a movement too swift for the eye to follow, and could also beat fore and hind wings independently.

During the past thirteen years I had not recorded a libellula earlier than 23 May. Indeed, despite a fiercely hot week before mid-May the previous year, with the thermometer registering between 60° and 70° F, not one was to be found before the 25th; but the first week in May that year had been very cold, and the factor influencing the emergence of these spring dragonflies was perhaps the temperature of the water in the larvae's pool, or possibly the amount of direct sunlight that had fallen upon it in preceding days. Only when the water in their pool had been heated to a certain degree were the nymphs stimulated to break through the water-barrier and clamber up the reed stems to attempt − not all succeeded − their fantastic metamorphosis: unfolding their winged beauty from the ruptured shell of that hideous aquatic armoured case, which would be left clinging firmly to three reeds by rigid, hollow legs! If, on first leaving the pool, the nymphs encountered an air temperature below about 50° F those that had not already begun to rupture their larval cases returned to the water. They would make further efforts to metamorphose the following day or days, though clearly there was a limit to the time this climacteric could be delayed.

However that may be, the precocious 10 May libellula was shy and not yet strong on its immature 'glassy' wings, seeking the shelter of the heather tangle and not allowing me to approach closely, foiling all such attempts by short, relatively slow flights of twenty or thirty yards from one resting place to another. But a week later, when it was still scorching hot, the Milton peat-moss was the scene of much activity. There had been a strong southerly breeze the previous day, with the result that the few libellulas present at the moss had been quiescent for the most part, only occasionally making brief flights over the small weedy pool (some thirty yards long by twenty-five yards across) impounded in the heather. The following morning, however, a dozen at a time were flighting over the pool, and mounting up twenty feet to rise above the small pines dotted about the moss — magnificent bronze-gold insects in the fierce sun, their wings now suffused with saffron. There was a gleaming golden sheen on the hairy abdomens of those at rest with wings spread, while clinging with their middle and hind pairs of legs to reed stems and cleaning their mandibles with the front pair; but they were still immature, and the rich red-gold bloom would soon fade.

At the pool each male patrolled its special beat, and there were frequent rustlings when two clashed at the limits of their respective territories. Every other second one would interrupt its patrol flight to hover briefly, now here, now there. A quarter of a male libellula's flying time must be passed in these intermittent hoverings on swiftly fanning wings. What purpose did they serve? They did not constitute hunting stations, for though the pool and its surrounds swarmed with insects only exceptionally was an attempt made to capture any. Undoubtedly they were watching stations for trespassing males and for females flighting in to the pool to lay their eggs. Since the latter was an infrequent occurrence at this date, the males had the pool to themselves for the greater part of the day. Thus I had been waiting for a long time before a gravid female suddenly arrived at 10.15 a.m. and was immediately seized by one of the two males present at the pool at that time. Coupling, mating and disengagement were almost one instantaneous act in their savage, yet technically marvellous aerial pairing; for in what was virtually a single complicated manoeuvre the male gripped the female by the 'nape' with his abdominal claspers, while she looped her abdomen down and then curved it round and upward in order to attach her pairing apparatus (placed beneath the eighth segment) to the accessory genitalia beneath the second segment of the male's abdomen. His genitals were in fact located on the last but one segment, but, by bending his body, he transferred sperm to the pairing organs on the second segment.

On being released by the male, the female began to lay her eggs. Lashing the water violently with dipping abdomen, she expelled a single white elliptical egg (one-sixteenth of an inch long) into the water at intervals of six or twelve inches around the edge of the pool, while hovering briefly over each laying station with wings vibrating as rapidly as those of the hummingbird hawkmoth poised above its flower, and darting from one station to the next. The majority of the eggs sank to the bottom, as did those of most other dragonflies that broadcast them in this manner. From time to time the male returned to hover near her, and when she was about halfway through her series of between twenty and twenty-five layings there was a second swift aerial mating.

Contrasting with the libellulas were the red damsel-flies (*Pyrrhosoma nymphula*), which had the most extensive local range in Badenoch of any of the smaller dragonflies. A few indeed might be found at the scattered peaty pools nearly 1,500 feet up on those remote moors around Ben Alder. During the previous day's strong breeze numbers of the immature bronze-green and dull red males with glassy wings had been sheltering in the thick tangle of heather in the peat-moss; but among the scores fluttering up from the heather this morning were a few fully adult males, distinguished by alternating stripes of metallic bronze-green and orange-yellow on the sides of the thorax, and with gleaming blood-red rubies for eyes. Many, including the immature, were coupled in tandem with females, which differed from them only in being a duller red, and having yellow instead of bronze-green abdominal rings. In fact eight or nine out of every ten coupled pairs engaged in laying eggs were immaturely coloured, and unmated females were also to be seen taking up laying stations on the leaves of bog-bean. One male, gripping a female by the neck in the normal pairing position, alighted on a leaf floating on the pool. His partner then looped down her abdomen into the water, articulating each segment separately in the manner of a 'stick' caterpillar, and probed tentatively here and there with it before finally selecting a suitable place in which to lay her eggs. Then for several minutes she was employed in attaching them to the underside of the leaf, though without any apparent movement of her abdomen. Judging by the duration of this operation she must have laid a large number of eggs around each leaf; but although she was ultimately almost completely submerged, the male remained coupled to her throughout. In the initial stages he was balancing solely on her neck, retaining his precarious upright posture by fanning his wings intermittently. Eventually, however, a breeze forced him to lower himself into a horizontal position along the leaf, until she had finished laying, though he resumed the upright posture again at her next laying station. When he finally released her, after the two of them had rested for some time on a reed, they moored themselves to different stems, abdomens projecting at right-angles.

It was dull and cold at the end of May with night frosts and temperatures in the forties by day. Not a dragonfly was to be found anywhere. But on the morning of 6 June, when it was twenty or thirty degrees warmer, the libellulas were again on the wing in numbers over pool and heather and pinewood peat-moss, and a few were resting with all four wings spread while clinging to the pine boles. At nine o'clock twos and threes of patrolling males were chasing and clashing around the pool in the peat-moss; but whenever a cloud crossed the sun they alighted and disappeared, for their body temperature approximated to that of their environment: when the sun shone they were warm, in cloudy weather and at night they were cold. Shortly after ten o'clock a coupled pair swooped in to the pool − only to uncouple immediately, when the female was at once pursued by other males. Finally another female − or possibly the same insect − during the course of laying about sixty eggs at various stations, was seized and mated on no fewer than eight separate occasions by at least three different males. With fifteen or twenty males at the pool at one time, the three or four females arriving between ten o'clock and ten-fifteen were repeatedly seized by one male after another. Most couplings lasted for no more than five seconds before the pair broke apart, usually while in flight, though one or two pairs alighted to do so. A few, however, flew round the pool in tandem for some minutes, with the males providing the motive power and towing the glider female. Such couples might be pursued round and round the pool by unattached males. If chased a short distance away from it, they would invariably return to it, when the female, on being released, would immediately zoom off to begin laying at various stations. Thus the pool was primarily a mating and laying resort.

June 11 was a day of hot sun and showers. Scarlet damsel-flies darted up and down sunny rides through the woods, and on a heathery bank, gay with spotted orchises (both pink and white), at the edge of the peat-moss were a scuttling lizard, a burnished-copper fritillary butterfly and the first great golden-ring dragonfly (*Corduleg-*

aster boltonii) devouring a small beetle with an audible brittle champing of mandibles. Since dragonflies were oblivious to all else when feeding I was able to admire at leisure the magnificence of this brown-eyed immature specimen: pale amber face, black and yellow-gold head and thorax, and black abdomen (three inches long) with irregularly spaced, bright yellow rings. The beetle disposed of, it rose from the heather and settled thirty feet up a pine.

The following morning the temperature was nearly 70° F, and at nine-twenty, after I had waited at the peat-moss for half an hour, an ivory-faced female libellula came in to lay six eggs in a runnel. She was then seized and coupled for seven seconds by an immature male with a white 'fault' across the tip of one glassy wing, who towed her to a nearby muddy pool in which she laid a further nineteen eggs, while he hovered over her. When she made off after laying, he took up his stance at the edge of the pool and assaulted a pair of red damsel-flies, the female of which was preparing to lay. He assaulted them again after she had laid in the mud, which took her about four minutes, and also attacked but failed to capture two moths. Damsel-flies, laying in pairs and flying in tandem, were everywhere in the moss; and in one instance the curious formation of a tandem of three was retained both at rest and in flight, with the female in the rear vainly trying to mate with the second male. Four pairs, prospecting laying places in the tiny pool, were repeatedly threatened by the gigantic libellula, though he did not actually savage them or even drive them away, and when I left the pool at ten o'clock he was still at his station.

At an adjacent little pool there were two male and three gravid female libellulas. One of the latter was seized after laying thirteen eggs, and a second after laying six, while the third, after being coupled, laid thirty-four eggs. But though the male hovered over the latter all the time that she was laying, he strangely ignored her when she rose to capture a moth and settle fifteen feet up a pine. Then, after all the females had left the pool, the resident males took up their familiar territorial stations, either while hovering or on reed look-outs. From these they would sally out in swift chases with trespassing males, for rather than one pursuing another, the two appeared to accompany each other. Two males, for instance, would whir round and round the pool together and as far as twenty-five yards away from it, mounting perhaps to a height of thirty feet; only to return immediately (often together!) on a direct course to the pool — back again before I could turn my head. Their territorial points of vantage were spaced from twelve to eighteen feet apart, and any intruder approaching within eight feet of these was attacked. As roving males, and also damsel-flies, passed, so the occupants of territories rose one after another to take up the assault.

By ten forty-five there was intense activity at the main pool, with twenty and at times thirty libellulas present and coupling, and there was a constant rustling of clashing wings; but when I returned in the afternoon it was dull and breezy though still sultry, and the males were behaving differently. Instead of returning to their vantage points after every sally, as they had been doing in the morning, they hovered persistently, apparently feeding. Nevertheless, every now and again two would chase away thirty yards from the pool with a birdlike swoop of wings past my face; and almost every 'pair' would follow the same course round one particular pine tree and, as in the morning, be back again at the pool before I could turn my head.

On the afternoon of 22 June it was at first dull and sultry, and only one and subsequently two libellulas were at the pool. One was making regular circuits of it, though halting abruptly in swift flight every few yards in order to hover briefly, and returning again and again to its reed-stem lookout. But as the sun began to break more strongly through the cloud, so their numbers increased from five to ten and finally to seventeen, with six at a time darting in to the pool from their resting places in the heather; and once again I observed the phenomenon of males constantly leaving the pool in pairs and chasing around that particular pine tree. Included among the newcomers were some females, which were inevitably immediately seized by males; but coupled pairs were making much longer flights than previously, remaining in tandem for ten or sixteen seconds before separating quite gently and unhurriedly and alighting in the herbage. One female, after laying eighteen or more eggs, was seized and towed round the pool before being released. When she attempted to resume laying she was seized again, and the pair, while looping the loop, mounted to a height of forty feet before breaking apart. A second female, after a brief coupling, laid more than one hundred eggs while the male hovered above her, before she ultimately rose and flew away from the pool.

The following day I was at the Badan Dhu crags on the 1,200-foot level of the moor above Drumguish. There, at the little pool almost surrounded by pinewoods at the base of the crags, were not only the inevitable twos and threes of chasing libellulas, but the first tiny, brilliantly painted, blue damsel-flies (*Enallagma cyathigerum*): the males a shimmering sky-blue, the females of paired couples a very pale grey-green. In some years they emerged as early as the first week in June, and when I returned to the crags on 2 July I found them shoaling over the pool in scores, with males outnumbering females by twenty or thirty to one. They were also present at a larger pool, deep enough for my spaniel to swim across, on the open moor. Many were hunting minute insects, though that seemed too crude a term to employ in relation to such fragile creatures, whose wings were invisible from most angles. It was, indeed, their shadows on the sunlit water that first arrested my eye. But despite their agility they were as aggressive as other dragonflies, nosing at the whirligig beetles and even darting at the three or four huge libellulas whirring around the pool at terrific speed. From time to time they would cease hunting and attach themselves to blades of water-weed. Three or four might be moored to a single blade one above the other, when, with their wings folded down their abdomens, which projected at right-angles, they resembled miniature airships at a mooring-mast. Here and there pairs were alighting in tandem on the weed, and the female of one pair, after looping herself almost into a circle, dipped her abdomen into the water and then submerged for some minutes, while the male rested on the weed. However, most pairs attracted so much attention from unmated males that the females were obliged to lay their eggs very expeditiously on streamers of weed just below the surface, though this did not prevent their partners, still in tandem, from maintaining an erect posture by fanning their wings with maximum velocity. Such a female might be the centre of attraction to four males, all attempting to attach their claspers to her neck, despite her being in tandem with a fifth; and in at least one instance an unmated male succeeded in knocking off the coupled male and obtaining a hold on the female, whose head was below water and her abdomen high in the air! The females took such tremendous punishment, being forced under the surface or dragged on their backs through the water, that it seemed inevitable that their wings must either become waterlogged or be torn to pieces; and I was astonished to see that they could survive the roughest treatment without apparent damage and take to the air still in tandem.

12 The beautiful Aeshnas

THE LIFE-SPAN of the individual libellula that survived the critical first few days after emerging from the larval pool was probably no more than one or at most two months, and early in July the libellula year was drawing to a close. On the sultry afternoon of the 10th there were no females at the Milton pool and at no time more than nine males. Even when the sun broke through the cloud the latter were slow to appear.

Their place was filled by the beautiful blue aeshnas (*Aeshna caerulea*). Although I had been seeing occasional solitary specimens of this Strathspey rarity since early in June, it was the 28th before numbers — the vast majority males — were on the wing in the Milton moss and pinewood rides, hawking high in the wooded part of the moss and soaring up to the treetops after their prey. One, flying low, would circle persistently round and round a beat of a few square yards along a pinewood ride, intermittently gliding with all four wings spread, and occasionally mounting thirty feet after an insect, or to a height of sixty feet when a breeze blew up. The pinewood edge was their special haunt, and wherever there was a stand of pines, whether far up the Corrieyairack or isolated on the Ben Alder moors, there one could be certain of finding one or more blue aeshnas.

Males, females and immatures presented a most puzzling variety of colour patterns. To the artist's eye the male was predominantly a splendid blue, pricked out in black, with three pale yellow wave-lines zigzagging across the rich, dark velvet-brown thorax, while his wings were an extraordinary dark copper. The female's colour pattern, though not so bold as the male's, formed an exquisitely delicate mosaic composed of galaxies of blue-green dots and peardrops; her eyes were lilac-brown in contrast to the blue of the male's. The immatures displayed an almost limitless range of patterns and colours, with off-white stripes on the thorax, bluish-white or yellow-green mosaic-work and yellow or white spots on black or chocolate-brown abdomens, and white zigzags along the sides, and in some individuals a bluish-white tip to the abdomen.

94

When in the early evening of 14 July I came down Carn Ban, after a day on the high tops, to that ever-welcome pinewood with its snow-cold waterslide in Glen Feshie, solitary blue aeshnas or pairs were sailing high and low between the pines. The steep winding path through the wood led me to and from the hawking beat of one solitary, which did not appear to be normally coloured. As it passed and repassed at head height within a foot or two of me, with wings glinting coppery in shafts of sunlight filtering through the trees, I finally realised incredulously that this magnificent specimen was exclusively black and white, for its black abdomen bore a pattern of white peardrops, and the black thorax was boldly striped with three white bars. The lilac-brown colour of its eyes indicated that it was an immature. I had seldom been more excited. Was I the first naturalist in Britain (or in Europe for that matter) to be privileged to observe a 'white' aeshna?

The following afternoon numbers of aeshnas, including some of the white variety, were hawking their separate beats in the Milton pinewood and moss: now at head height along a ride, then darting up with unbelievable swiftness to the pine tops to capture a small moth or lacewing in the basket formed by their projecting, bristly legs, and whirring down again to chew up their prey at leisure while clinging to a pine bole. With long straight wings set well forward on long abdomens they resembled model biplanes as they glided about the treetops. From time to time a female would hawk into a male's beat, and there would be a brief coupling on the wing; but there was little aggressive rivalry when two or even three males met.

By the windless afternoon of the 25th the Milton colony of aeshnas was at its optimum strength of perhaps a dozen individuals at the pinewood edge. As usual there were none at the pool where, however, the first gleaming molten-bronze, black-legged sympetrum (*Sympetrum nigrescens*) was on the wing. The adult aeshnas were easier to approach than the immature, and I was able to study at a distance of only twelve inches a superlatively beautiful male with an almost solid-blue abdomen. The delicate black patterning on the latter contrasted with the blue and brown thorax, the curiously 'glassy' sides of which were striped with faint whitish lines. At such close range the blue eyes appeared black above the pale greenish-white mask.

This insect was chewing a lacewing, whose wings it ultimately rejected. The larger hawker and darter dragonflies were most uninsectlike and unpredictable in their reactions to the human intruder. At times they were shy of him, at other times they ignored his presence, particularly when engaged in masticating their prey. They were always acutely alert visually, and apparently also aurally, for when on one occasion I was stalking an immature aeshna, resting on the herbage, very cautiously from the rear, and had got to within nine feet of it, I snapped a twig with my foot and it flashed away, zooming up above the trees.

With temperatures in the eighties night and day, and the next morning the warmest I could remember on the hills, a golden-ring was hawking successfully over the heather at an altitude of over 2,000 feet. But although 'white' aeshnas were still numerous, blue adults had almost disappeared, and were being replaced by the first common aeshnas (*Aeshna juncea*). On 5 August — the seventh consecutive fine warm day — both male and female *junceas* were at the Bailleguish pool, 1,000 feet up on the moors above Drumguish. Now and again a coupled pair would loop the loop, when the contrast between the dark nigger-brown and greenish-yellow female with the bright blue and black male was startling. There was an apple-green tint in the male *juncea*'s blue that always distinguished it from the pure Cambridge blue of the *caerulea* male, and the *junceas* were further distinguished by the broad yellow or green and yellow bands across the side of the thorax. Although they were bulkier, with a $3\frac{1}{2}$ inch wing-span, size in the field was not a safe distinction between the two species, except in the case of many unusually large *juncea* females, with typically decurved abdomens, which could never be mistaken for the aristocratically slender *caerulea* females. There were some years when numbers of these immense *juncea* females were abnormally coloured, with yellowish-brown abdomens bearing a white pattern faintly tinted with blue, and curved white bands across the side of the thorax. So predominant was the yellow in their colouring that at a distance they appeared to be yellow and brown dragonflies.

Individual *junceas* were flying low over the pool (which I could not recall ever seeing the blue aeshnas do in the peat-moss), and the brittle clashing of their wings as they turned and accelerated was as regular as it had been among the libellulas earlier in the summer. As they hawked to and fro along the edges of the pool, visiting every little inlet, with a slow 'ferreting' flight, their gauzy wings rustled continually. Was this rustling contrived deliberately for the purpose of flushing insect prey? Every now and again one male would make a pass at another or at a posse of small black sympetrums (*Sympetrum danae*), which were on the wing for the first time, rising in 'clouds' from the heather at the edge of the pool. The confidence of some of

1. A Roe Deer Buck on the moors

2. A Red Deer Stag on the moors

3. (Left) A Golden Eagle perched on a crag
4. (Below) A Buzzard landing at a Rabbit
5. (Right) A short-eared Owl perched on the heather
6. (Bottom right) A Peregrine Falcon

7. (Left) A Whooper Swan
8. (Bottom) Lapwings
9. (Below) A Common Sandpiper on a river boulder
10. (Right) A Curlew in flight

11. A pair of Scottish Crossbills:
courtship feeding

12. A pair of Ring-ousels at
their nest

13. A Mountain Hare

14. A Wood-ant hill

15. A Common Aeshna
Dragonfly

16. A Bumble Bee on sea holly

the *junceas* contrasted with the shyness of most of the blue aeshnas, for while some sheered off in alarm when flighting close to me, two actually alighted on my chest.

Still more delightfully confident were the black sympetrum males, which would come one after another to sun themselves on the back of my hand. In these early days on the wing the females — distinguished by brilliant yellow thorax stripes — and immatures outnumbered the adult black males by at least twenty-five to one. The hairy little males, which maintained an almost perpetual hover-flight, from which they would make sudden darts forward (or backward!), curiously resembled small fishes, not only in their abrupt movements but also in appearance because of their extraordinarily swollen thoraxes, which contrasted with the slender lines of the damsel-flies, though there was considerable individual variation in size among the sympetrums. The few pairs in tandem were even more like fish, as they constantly danced or, rather, bumped up and down in one place while the female dipped eggs on the surface of the water and weeds. During this operation unattached males nosed around inquisitively in a fish-like manner. On one pair uncoupling after the female had finished laying, the latter was immediately knocked down again and seized by a male — possibly by her original mate.

After prolonged observation at this moor pool I was delighted to distinguish a new species of dragonfly — new, that was, to me. It was so unbelievably fragile that I could be excused for having overlooked it previously, when paying special attention to the more interesting and obvious behaviour of the larger hawkers and darters. The newcomer was the green lestes (*Lestes sponsa*), three or four individuals of which were making short weak flights from one weed-stem to another at the edge of the pool. It was an exquisite little insect with a green face and dull blue or brown eyes, palest blue head, powder-blue thorax with a scintillating orange-red or chrome or copper-red 'jewel' set in its side, and gleaming metallic-green sides to the brown abdomen, of which the tail and adjacent segments were a powder-white blue. It was curious that such delicate insects as these and the blue damsel-flies (two or three males of which were also present) should select as one of their habitats the most exposed moorland pools, whose waters were seldom calm, with the result that it was only when they were mirror-smooth that shoals of damsel-flies could venture out from the weedy edge and the few lestes embark upon their feeble cranefly-like flights from stem to stem. An occasional lestes might indeed be found at a temporary rain-pool on the braes. Yet, oddly enough, the longest known life-span of any European dragonfly was that of a marked lestes recaptured not less than ten weeks after metamorphosis.

On 11 August the lestes were laying in the Milton pool. One coupled pair were clinging to a reed stem: the male with wings folded, the straw-coloured female with all four wings spread like a windmill's sails. Subsequently the latter began looping down the stem until almost submerged, in order to lay her eggs below the water; then walked up the stem again, with the male still attached, to repeat the operation on another reed. Elsewhere three were in tandem on a reed, and after I had been watching them for three or four minutes the topmost male jerked violently once or twice and released himself from the second male; another solitary male was dragged beneath the water by a water-boatman, which was propelling itself with its two long flexible 'oars' just below the surface.

It was now sixteen days since I had seen any blue aeshnas. Another season had therefore ended without my being able to discover where they laid their eggs. In view of the hours I had devoted to watching them my failure to do so was puzzling, though I suspected that they laid in damp 'pans' in mosses containing little or no water. However, on the afternoon of the 11th, *juncea* aeshnas were behaving very strangely, for single brilliantly blue specimens with yellow and pale green stripes across the sides of their brown thoraxes − and therefore unquestionably *juncea* males − were flighting to and fro some nine inches above the pool and apparently broadcasting eggs at random from decurved abdomens with audible rustling clicks! Fortunately for my sanity I did not actually *see* any eggs, and I was later to learn that the males of some other species of dragonflies had the inexplicable habit of dipping their abdomens in the water while flighting over it, as if ejecting eggs. Actually, although the *juncea* males appeared to be going through the exact motions of laying, the Aeshnidae did not in fact broadcast eggs like the Libellulidae, but were equipped with serrated ovipositors with which to insert single eggs into weed or mud.

All the *junceas* in the peat-moss were concentrated at the pool, but in the glen below the pinewoods I came upon one of these beauties basking in the sun while clinging to the trunk of a birch. For a long time I stood looking into the exquisite clouded, yet crystal depths of its huge, pale blue globular eyes, compounded of several thousand facets, wondering what lay behind the sphinx-like mask of that black-lined, yellow face; and I noted down a detailed description of the

marvellous colour-pattern: thorax chocolate-brown with two narrow
yellow stripes on the upper part and two broad yellow slashes on the
sides; two clusters of blue spots and lobes at the wing-attachments,
together with two large blue lobes separated from two chocolate-
coloured lobes by two small yellow triangles; abdomen black, rather
than brown, with nine sets of twin blue lobes and five intermediate
sets of small yellow twin triangles on the fore three-quarters, with
single blue lobes along the sides opposite to the yellow triangles.

Also in the glen, clinging to a rush stem — wings in perpetual
motion — was a magnificent golden-ring with enamel-green eyes and
broad green and yellow stripes on the sides of the thorax. What a
solitary wanderer this dragonfly was. In all my years in Badenoch I
never saw two together, and only once did I observe a female laying in
a burn at an altitude of 1,000 feet: for a golden-ring's method of
laying was to force her eggs into the sand or gravel of a shallow,
sheltered recess in the bank of a hill stream by means of a unique
prolongation of her ovipositor sheaths, which formed a hard black
projection to her three-inch-long abdomen.

Throughout August there was no rain. For thirty days the sun
shone from dawn to dusk and, with temperatures in the seventies and
eighties, it was the hottest month within memory. Aeshnas were
constantly on the wing over garden, pools and river, sailing over the
moors and hawking their erratic beats up to the headwaters of the
glens, up the rocky burns for 2,000 feet into the hills, and even on the
high tops above 3,000 feet. Wherever there was a lone stand of pines
on the remotest moor, there were the blue male *junceas*, like the blue
aeshnas before them. They only among dragonflies were on the wing
in all weathers and even on the high moors when half a gale was
blowing.

Since the males were so commonly distributed it was curious how
seldom one encountered females; but on the 24th three or four very
large and splendid females of the yellow variety were laying in the
Milton pool. Attracted in the first place by a violent rustling in the
smaller of the two pools, I unwittingly disturbed them and, to my
chagrin, they made off. So yellow were they that I mistook them
momentarily for golden-rings, for all were yellow and brown without
any blue markings; even their eyes were brownish-green. Later,
however, I was delighted to find one of them again at the larger pool
and was relieved, after my experience on the 11th, to observe that she
was well down in the sedge and rush in the more densely herbaged
part, with half her abdomen submerged and probing among the

stones on the bottom. The violent rustling of wings that had initially attracted my attention was caused by the difficulty she had in forcing her way down through the jungle of stems. She remained for five minutes at this station, with her front pair of wings arched and vibrating with a mother-of-pearl scintillation while she laid, before moving on a little way. After she had been laying for twenty minutes at the second station I disturbed her, whereupon she zoomed away with that dead-straight flight so characteristic of the males. None of the latter was present at the pool at first, though six or seven were hunting in the sheltered part of the moss. They included a 'white' specimen with glittering coppery wings – the first *juncea* of this variety I had seen. However, after the female had been laying for some time, one of the hunting males did visit the pool, though he took no notice of her – if he saw her; but later he intercepted a passing female and knocked her down to the ground from a height of fifteen feet with the customary swift aerial seizure. The two then mated and uncoupled three times while looping the loop, before finally sailing up twenty-five feet and alighting on a spray of pine.

Also at the pool were nine black-legged sympetrums which, in contrast to the black sympetrums, were unsociable little insects. I had still not seen any mature red males – despite the fact that they had now been on the wing for a full month – though chrome-coloured individuals gleamed a molten red-gold when on the wing in sunlight, for there were copper-coloured veins in their wings and small saffron spots at the wing-attachments.

Between one-fifteen and three o'clock there was a north-east to south-west migration of large cabbage-white butterflies on a broad front across the peat-moss and moor, where those in exposed places were being blown west off-course. Apparently exclusively males, they passed through the moss mainly at head height – though ones and twos and half-dozens alighted on the scabious flowers – before mounting up over the barrier of pine trees ringing the moss. I counted exactly three hundred during this $1\frac{3}{4}$-hour period, before their passage terminated suddenly when the sky became totally overcast and the breeze freshened.

The first fortnight of September was mainly wet. Nevertheless on the 3rd, which was a very humid day with frequent showers and only occasional glimpses of sun, I found three newly emerged black-legged sympetrums climbing tentatively up their reed stem from the Bailleguish pool. Their 'tinfoil-plated' wings were either folded or just beginning to spread, and at this stage they had brown eyes and pale green faces, two brilliant mustard-yellow bands across the side of the thorax, and pale brown abdomens.

Nor did intermittent cloud and a stiff cool breeze on the afternoon of the 15th deter upwards of a dozen black sympetrums from mating at this pool. One male seized a female at a height of ten feet, and the two twirled over and over while falling to the heather. There the male released her, but, when she rose a minute or two later, recaptured her. Again, however, they fell apart, and only at a third attempt was the mating position achieved. After a short flight the pair alighted in the heather.

About five minutes after the sun had broken through the cloud a small, very pale-coloured and not fully mature male aeshna came into the pool. He was followed two minutes later by an equally pale female — she may indeed have been a 'white' specimen — who went to lay in a clump of weed in deep water in mid-pool, plunging her abdomen in almost up to her thorax. The male, though often hawking over her, took no apparent notice of her; nor indeed was he continually present at the pool, visiting it only at intervals after short or long periods of absence. At each visit he would whir round and round, only a few inches above the water, while hunting with rustling wings and being attacked from time to time by a pugnacious sympetrum. However, while it was true that a dragonfly's attention was particularly attracted by flying objects, whether other dragonflies or insect prey, subsequent events showed that he was aware of her presence. In the meantime she had stayed for only two minutes at her first laying station, but at a second remained for *fifty-three* minutes, despite the strong breeze. Throughout this period she was at times so deeply submerged that her hinder pair of wings actually lay on the water. At other times only the tip of her abdomen was under water as she probed about with it, while pivoting a quarter-circle or moving her position by an inch or two.

It was when she rose at the conclusion of this long session that the male, making one of his occasional visits to the pool, seized her in a most unorthodox way, for instead of attaching his abdominal claspers to her neck, he gripped her head with his mandibles, and thirty seconds elapsed before he obtained a normal hold with his claspers. This incident suggested the unlikely possibility that immature dragonflies might have to learn the mating technique, or did this one momentarily mistake the female for an insect prey? Whatever the reasons for this aberration, the male, having coupled with the female, attempted to fly up, but she clung obstinately to a reed stem. So, after a further sixty seconds, he released her: whereupon she immediately flew to another clump of weed, two feet from her original station, and began laying again. However, after only five minutes at this station she removed to another, but left this one almost immediately for a fifth, where she remained for nine minutes. At her sixth station she stayed only two minutes, and from her seventh (close to me) I unfortunately disturbed her, deeply submerged, when I was endeavouring to see if she was in fact a 'white' specimen, for some yellow females were so 'wishy-washy' in colour that they could easily be mistaken for 'whites' at a casual glance. After a total of seventy minutes' laying she then sped away over the heather alone, and did not return.

Five days later a pale yellow female (perhaps the same insect) was again laying in the Bailleguish pool. She laid at five different stations during a stay of twenty minutes, before flying away to hunt. Although a male passed over the pool once, he was the only other aeshna present. However, despite the continuing cool and mainly cloudy, breezy weather, with temperatures down in the fifties, there were a score of black sympetrums at the pool and in the heather, and the female of one pair in tandem was broadcasting what appeared to be single eggs at intervals right round the pool, ejecting each egg with a quick dip down to the water. When the sun came out for a rare hour numbers of the jet-black males would cluster on the sunny face of a stone dyke, and two or three at a time would chase one another off a warm stone, returning to it again and again; for unlike the females, which could be watched at leisure, the males were restless creatures, settling for barely sixty seconds at a time before darting up. Could a sympetrum see as well with its *ocelli* — those tiny flat and polished yellow-white discs at the top of its head — as with its true compound eyes? When I approached the males from behind they would rise immediately: yet they could hardly have obtained a rear view of me with their true eyes.

Both male and female black sympetrums, and male aeshnas too, would still be on the wing in the middle of October, and sometimes later than that, for hoar-frosts, severe enough to shrivel potato shaws and blacken broad-beans, did not kill off either of these species.

13 The Dance of the Whooper Swans

THE SPEY, when in spate, overflowed on to the adjacent water-meadows and rough grazings through huge demolitions in its banks, which had originally been excavated by rabbits, and then enlarged by sheep and cattle seeking relief from flies and midges. After periods of heavy rain, and when the snow melted on the hills during the spring, the meadows and marshes became an almost unbroken sea of waters for mile after mile. Only the banks of the breached dykes and a few ancient alders protruded above the floods.

Although a few hundred cattle and sheep still grazed the Spey levels, Nature had taken over from man the agriculturalist, and they were now predominantly the haunt of duck and wild swans during the winter, and the nesting grounds of many kinds of wading birds and a couple of thousand black-headed gulls in summer. The sedge warblers, magpies and pheasants that nested in and around the marshes and their wooded fringes were not to be found anywhere else on my beat. In the spring roe deer and hares browsed the green rushes among the gleaming clumps of golden saxifrage and bronze-gold kingcups. Later there would be pale globe-flowers, yellow-rattle, the glowing magenta spikes of marsh orchises and, crowning glory, a massed acre of the purple and ultramarine obelisks of lupins. Their dark greenery almost concealed a dense growth of woody cranesbill, whose delicately shaded clusters of pale mauve flowers were dusted with the purple of a butterfly's wing. At high summer the grey-green water-meadows were slashed and ringed with snowy swathes and clumps of bog-cotton.

Throughout the winter families of whooper swans and small herds of adults were constantly coming and going between the Spey and almost every loch, large and small, in Badenoch. By day and by night their musical bugling broke the winter silence, paradoxically recalling summer with its likeness to the cuckoo's calls. The smaller lochs froze over very quickly, and also unfroze with disconcerting rapidity; but when they were ice-free, and sometimes when they were not, there were few days throughout the winter until the end of April when there were not swans on the twin lochans below Craig Dhu. However, they did not usually visit the small lochs at some distance from the Spey when they first arrived in Badenoch. I was therefore surprised one year to find the earliest swans of the autumn feeding on these lochans as early as 4 October. Never before had I seen a family group before the 17th, and there were two unusual features about this family. In the first place it comprised three cygnets accompanied by only one parent — which was exceptional; and in the second place the cygnets were the youngest I had ever seen, with just the faintest tinge of pink on their grey beaks. One could almost discern the down on their soft smoky-grey necks. Moreover they could only utter a barely audible imitation of their parents' trumpeting. Were they perhaps Scottish-bred cygnets? Even in Badenoch individual swans occasionally prolonged their stay until the middle of June.

That was only the fifth occasion in seventeen years that the whoopers had returned to Badenoch as early as the first week in October, and on three of those occasions they had been migrating through the strath, for the earliest comers were normally on passage in herds of up to nine. Although they might go through at any time of the day they did so most often before nine in the morning or after dusk, and one could hear them passing, always bugling, as late as nine at night. If they did not return in the first week of October, then they would not do so until after the middle of the month when, quite often, their arrival coincided with the first heavy snowfalls on the hills. On a still day, when windless forest and ice-bound river were hushed in the solemn white grandeur of the mountains, I would hear the swans conversing musically on the marshes. At dusk more would come lifting like ghost-moths into the dead-white hills, to glide down to a mere still partially free of ice, letting fall singly their melodious bugle-notes. Just before darkness fell there would be a tumultuous outcry from them, but thereafter only an occasional soft, sleepy grunt.

There were years when no swans arrived in Badenoch before November, and it was always that month before sizable herds settled in to winter on the Spey marshes and their historic haunt of Loch Insh, where a herd of from seventy to one hundred had been referred to as habitually wintering as far back as 1792. It was most noticeable how, within the herd, family groups, pairs and single swans fed separately and flew from place to place in these discrete units, while those pairs with cygnets drove away other adults feeding in too close proximity to their cygnets. Exceptionally a herd might feed on undersown stubble at a distance of five hundred yards from the river and close beside the railway. They would indeed do so only two fields away from noisy potato-pickers — flooded potato fields provided an additional source of food — and on one occasion I surprised one individual walking up from the river to the distant stubble. My dogs found its scent almost as interesting as that of hares or ptarmigan.

At the Craig Dhu lochans, with the Rock's 1,000-foot wall of cliff as a gigantic sounding-board, the swans' beautiful bugling notes and their echo were amplified with Wagnerian grandeur. One March day there were three pairs of adults on the lochans. At my coming the three pairs swam towards each other and, on meeting, both birds of each pair rose up and beat their spread wings on the water, while sounding a fanfare of crag-resounding trumpeting; after which each cob and pen trumpeted to its mate, with dipping heads almost touching.

Many years earlier I had watched a more complex form of this unique ceremony on Loch Awe. I quote:

By a most happy chance, on a morning when there was a partial thaw after three weeks of intense frost, I had only just begun to fell some old thorn trees at the edge of the loch, when two whoopers and their cygnet came flighting in to their favourite feeding place with a clamorous and discordant whooping, so loud and persistent that it set the dogs barking at the cottage on the hill above the loch. Very shortly six more came in with the same clamorous whooping, to alight on splayed paddles, with high-prowed sternums churning up the hissing water and straight necks sloping back at an angle. At this, the pair with the cygnet began whooping again, and then tremulously waving their great pinions, stretching them to their full extent almost flush with the water, while raising and lowering them only slightly: gradually accelerating this delicate undulation and as gradually retarding it, simulating the dance of the dying swan. At one moment the two would display in this manner to each other, then to the cygnet, then to the newcomers. Then, with wings still waving, they danced on the water on their huge paddles, before suddenly rearing up breast to breast with a vehement dipping of proud head to head and a mighty beating of vast hollowed wings — all to the triumphant whooping. A final rising in the water to flap wings vigorously, and the ceremony was concluded: this unforgettable incident in a naturalist's life.

14 A Winter Concentration of Dippers

WHEN walking on the banks of the Spey west of Newtonmore one afternoon in the middle of November I was amazed to count no fewer than thirteen dippers — all but one in full adult plumage — positioned along the inside curve of a five-hundred yard bend in the river. Each occupied its own station — in the majority of cases a rounded boulder at the water's edge under the trees on the south side of the river — from twenty to forty yards distant from the next station. This was a phenomenon of which I had not previously been aware, since the hill streams of the high glens, rather than the Spey, had been my province.

Although the partially submerged boulders were the most favoured stations, the branch or root of an alder a foot or two above the water might also be used, and in one instance the occupant of a boulder station was flicking its wings aggressively and even leaping up at another dipper perched on a branch above it. That was an exceptional proximity, for though one bird might remain on its boulder when another alighted within five feet of it, and though two might take off from their separate boulders in a pursuit-flight every now and again, only once did two settle for a second on the same boulder. Normally, on being 'visited' by another dipper, the occupant of a boulder would quit it instantly, either to alight in midstream or to fly hundreds of yards up or down river and take up a new station, possibly supplanting a bird already in occupation without more ado than the inevitable series of bobs on alighting.

From time to time one would plop into the river from its boulder to fish briefly, or dart out after another flying past; and many were singing sweetly and loudly, either from their boulder stations or, characteristically, from a perch on a branch of an alder an inch or two above the water. From October and November on into March and April there were always dippers singing in the Highlands, especially at first light on those mornings when the strath was filled with drenching mist.

Their concentration in that one particular bend of the Spey on a reach of several miles could no doubt be attributed to the abundance of suitable boulders, both at the edge of the river and also in midstream, which served as convenient fishing perches, and to the considerable area of stony shallows which were ideal for surface-fishing.

By the first week in December the number of dippers occupying stations within the bend had increased to sixteen, and as many as six might be fishing at one time in the shallows — bobbing up intermittently — at that point where the river swept round at its greatest breadth. A dipper lay curiously low in the water when surface-fishing for floating prey, often whirling around like a phalarope. All its movements indeed were those of a jack-in-the-box. At one moment it was on its low perch above the water: the next instant it had vanished, without my observing its departure. Twos and threes were continually flying aggressively at one another, both from boulders and from the water while fishing. Many were in song while in pursuit-flights across the thirty yard breadth of the river, and when one alighted after a chase it might 'explode' into a linked run of *zeet* notes.

Subsequently, torrential rains and melting snows engineered a full spate on the Spey. I suspected that, with fishing boulders submerged, this remarkable concentration of dippers would evacuate their stations and return to the various hill-streams on which they had nested, and where there were always a few to be seen throughout the winter at altitudes as high as 1,700 feet. There were, for example, ten or twelve nesting pairs in Glen Tromie — approximately one pair to every mile of river — from its juncture with the Spey to its source in the Forest of Gaick. Indeed pairs in song became commonly distributed on the hill streams in October and November. My supposition proved to be correct, for during the flood period only two stations were occupied, though a third bird was fishing in a pool on the adjacent marshes. But when the river dropped to normal level on 10 December there were twenty occupied stations along a stretch of eleven hundred yards that included the concentration in the bend. There may have been more than that, since it was extremely difficult to detect the dippers against the snow-capped boulders. When they were at their stations under the trees on dark midwinter days all that one could see of them were the snow-white discs of their breasts, until a ray of sunshine illuminated the rich, dark-chestnut band below the white.

At that time I had still not begun to understand the reasons for the constant, apparently aggressive pursuit-flights and the continued refusal of one dipper to remain on its boulder when another flew at it. However, after one 'pair' had engaged in a series of pursuit-flights, one of the two did alight on a boulder within three feet of the other: whereupon the latter, instead of vacating its boulder, threw up its head, straightened and spread its tail, and assumed an ousel-like posture with a splendid display of white throat and breast, while warbling a few notes. I presumed, therefore, that this was a cock displaying to a hen, and not to another cock. Previous pursuit-flights involving two, three or four dippers might thus have been sexual chases, though of an extremely aggressive nature.

Four days later twenty-one stations were occupied along a one-mile stretch of the river. I had always made my rounds in the afternoon, at which time the dippers were invariably fishing, bringing up larvae after almost every dive. When they saw me approaching, the majority would leave their midstream fishing and fly either to boulders near at hand or from fifty to a hundred yards up or down river and make a belly-landing in midstream again. Never would two alight on one boulder. If one did pitch on another's boulder, the latter inevitably left, as was also the case if one 'pancaked' near another fishing in midstream. At one point, however, four did occupy a six feet square of boulders for a few seconds.

On Christmas Day the Spey flooded over its embankment, and was still running full from bank to bank on the 28th, though eight feet lower. Nevertheless fifteen dippers were present shortly after noon that day. More were in song than on any previous occasion, and for the first time two tolerated one another on the same boulder for several minutes. Only twelve inches apart, they bobbed to each other alternately with precise timing: each genuflex being accompanied by a downward flick of half-spread wings from cocked tail. Yet this ceremony ended in apparent anger, for the two darted at each other aggressively both on the boulder and in the water, before ultimately taking off in a pursuit-flight, during which one was actually tweaking the other's tail at one stage; and a new feature of this pursuit-flight was that it included a number of soaring half-circles prior to the final long flight and total separation. I had always visualised a dipper's flight as being swift, direct and almost at water-level, but in fact it was slightly undulating with intermittent wing-closures, though the wings appeared to whir almost as rapidly as a wren's.

Despite the strong spate many were fishing, and one bird, flying out a little way from its boulder, took a curved header into the water with wings closed. A number of timed dives varied from three to seven seconds, with five seconds as the average 'long' dive; and after a five-second submergence the prey was invariably taken to a boulder for dissection. Midstream fishing was particularly favoured when the river was in flood, with a smooth 'oily' flow carrying debris churned up from the bottom. When the river was at normal level many birds fished among the stones at the edge, with only their heads immersed for one or two seconds.

It was now clear that every individual dipper — and I could still distinguish the odd immature bird among them — claimed one or more stations at the water's edge on the tree-lined, stony side of the river. One was actually below flood-level in the roots of an alder. But although territorially jealous of its station it would, nevertheless, vacate it — temporarily at any rate — rather than repel or share it with an intruder. The mere passing of another dipper, speeding swiftly by, was often sufficient to cause a station occupant to explode into a burst of song, developing from a reel of call-notes. Once in song there was no pause in the continuous flow of notes, which rose and fell in different pitches. On the swift Highland rivers, with their falls and rapids, one could not often distinguish the full tenor of a dipper's song above the perpetual noise of water rushing and tumbling; but on those occasions when one could listen to a bird singing beside some quiet pool, the song-pattern with its squeaky, grating counterpoint was not unlike the linnet's, though it also included short, high-pitched, lark-like whistles. It was indeed very varied and, at its best, impressed me with its resemblance to the songs of nightingale and the finest warblers, both in its throbbing utterance and in the diversity of its sprightly, throaty notes, which recalled the forceful *prrrr-ee* of Cetti's warbler.

On the afternoon of 5 January, when the Spey was partially frozen, the thirteen dippers present were pattering over the ice in pursuit of insects as nimbly as an attendant pied wagtail or a sanderling at the edge of the waves; or running along the edge of the ice-sheet and taking headers into leads of open water, to surface with the inevitable larvae. One actually swam under the ice-sheet, which was suspended on a stony shoal rising an inch or two above the water-level, and stayed beneath it for forty-five seconds. No doubt it obtained air from the pocket between the shoal and the ice ceiling. Another brought up a 1½-inch fish and, after pecking at it for a minute or two, gulped it whole with a considerable effort. At one time no fewer than eight were lined up on either side of a narrow, thirty-yard long, glacial-green lead, off which they dived so persistently, submerging for periods of up to eight seconds, that all were rarely on the ice at one time. One of these fishers curtsied to another a few feet from it, bowing almost to ice level with wings flicking at each curtsey, while the other remained motionless in that taut, upstretched, threatening posture already described; but inevitably both ultimately took off in the customary pursuit-flight.

By 11 January there were slight indications that these Spey dippers might be consorting in pairs, though ten or twenty yards still separated the majority of stations. They were, however, still aggressive, for on one bird alighting within a foot or two of another's boulder, it was immediately ejected by the occupant, instead of the latter vacating it in the customary 'Postman's Knock' procedure. On the other hand, when another dipper flew into a station in the roots of an alder occupied by a 'pair', both the latter flew off to take up separate stations fifty and a hundred yards distant from the alder; and it was possible that they in turn displaced other occupants. Thus, although I saw dippers in more or less the same places every day, and although it could not be doubted that each dipper had its own particular station, it was yet obvious that they were not territorially restricted in the usual sense of the term. Any dipper might leave its station and fly past other occupied stations for hundreds of yards up or down river; the longest such flight that I was able to follow from point to point extended for six hundred yards. Moreover, after observing what was presumably a pair standing within feet of each other and occasionally plopping in to fish, I was then surprised to note that a third dipper was fishing amicably in the same square yard of water, and that its boulder station was only twelve yards distant.

Some parts of the river still bore projecting arcs of thin ice, which were much favoured as fishing stations. Running along the edges of these ice platforms the dippers would take off in little flying dives (with legs looped together) or, alternately and repeatedly, dive in with a hop from a standing position, submerging for from three to six seconds, and bringing their numerous minute captures up on to the ice. When feeding in midstream, however, their prey was usually consumed afloat, as they spun around picking up stray dissected fragments; and not only would they bob up at the same spot that they had submerged, against a three- or four-knot current, but actually upstream of this. When not fishing or preening they would stand motionless for two minutes or more at a time, visible only as round white targets against the dark bank, for they bobbed only on alighting from a flight or when suspicious or alarmed.

Although the dippers' versatility included the proven, though inexplicable, ability to walk on the bottom of a river in water thirty inches deep for as long as thirty seconds, while turning over pebbles and gravel in search of crustacea, small molluscs, caddisfly larvae or dragonfly nymphs, they did not appear to do so on the Spey, where their maximum submergence was only eight seconds. Yet they had been observed doing so in fast-flowing mountain becks in water so shallow that their backs were often awash.

On 25 January the Spey, swollen with melting snow, and eighty yards across from bank to bank, was swirling seawards at a rate of seven knots through narrow reaches; but despite those conditions there were two interesting incidents. In one instance two dippers, which might have been presumed a pair, were frequently returning to close stations, and even to the same boulder, without apparent friction except for that aggressive upward-stretching posture. Yet, when they were subsequently fishing together in midstream, one suddenly flew at the other and a long, zigzag chase ensued, before they eventually alighted on either side of the river further upstream. Later, one of them — possibly the pursuer — flew to a third bird's boulder. After looking at each other for a minute, number three flew at number two and there was a brief scrap. Nevertheless, they later alighted near one another and reversed their roles, with number two flying at number three as a preliminary to a similar long, lifting zigzag chase. That ended with number three flying right out of the original 'pair's' reach, and number two breaking off the pursuit. In the meantime number one had continued fishing on the other side of the river, without taking the slightest notice of either the scrap or the pursuit-flight! I deduced from these incidents that number two was a cock and the other two hens.

In the second instance one of a 'pair' fishing flew at the other three times, with the latter evading these attacks by diving. Then there was a pursuit-flight, which included flying into and out of the water like teal in their spring mating engagements. The pursuer flew the faster, repeatedly overlapping the other, until they finally alighted close together. One then gradually made its way still closer to the other, until they were only three feet apart. And thus they stood, with one bobbing, for five minutes. Such mutual tolerance was unnatural!

Inevitably they came to blows again, actually fighting in the water, beak to beak, with threshing wings, before in the end breaking away up and down stream.

There had been very few afternoons during the winter that I had not gone my rounds of the dippers. Throughout January there had been an average of ten present every day, though the daily count varied slightly. This variation could no doubt be accounted for by station-holders flying out from either end of the control area, and perhaps by others being temporarily absent while visiting breeding territories on the hill streams; for while a few pairs nested on the Spey, none did so within the wintering reach. The day-to-day suitability of varying stretches of the reach for fishing might also have affected distribution, for if I counted many in the eastern reach there were likely to be few in the western reach, and vice-versa.

15 Dippers in Song and Displaying

On an afternoon early in February, when twelve dippers were at their stations, one flew across the river to alight within two feet of another, which reared up in song like a pouter-pigeon. This was the most mutually tolerant pair I had yet seen, for it was ten minutes before they began to bob intermittently, and for the most part alternately, on the stones at the edge of the water; though this led ultimately to one flying at the other and a short chase, before they broke apart.

Then, a scrimmage among three others resulted in a chase by two of them. On the latter alighting some yards apart, the hen of this pair drooped her wings and began to shiver them with such rapidity that their motion was blurred; and this she continued to do for some minutes, while the cock, advancing towards her from rock to rock, popped some morsel into her beak.

By 21 February I was convinced that these twelve dippers were in pairs. But no sooner had I come to this conclusion than one bird assaulted another, as inevitable preliminary to inevitable pursuit-flight; and subsequently four birds from different stations were involved in a chase. However, such aggressive behaviour between mated pairs would have been acceptable — with robins as exemplars — had it not been that so often after a chase the pursuer would break away to enact an identical performance with a third dipper. A presumed cock, for instance, was displaying to a presumed hen perched a little above him; but when a third dipper displaced the hen, the cock appeared to be indifferent to the change, though this new 'pair' subsequently chased in song, which included a loud squealing note that I had not heard before.

More convincing evidence of pairing was provided by two birds facing each other three feet apart on two big boulders. One, with tail spread, stood very upright — tilted back indeed from the perpendicular — while nictating continuously and singing very

loudly. The other, also taut, though less contorted, bobbed ceaselessly. After two minutes of this mutual posturing the latter — a
recognisable hen — flew to the base of the cock's boulder and bent
back her head so that she could look up at him, when both presented a
prominent expanse of white damask fronts. For a minute they stood
thus, and then *she* flew at him, and there followed the familiar
aggressive flying at one another, into and out of the water, before she,
again, flew to a stone and struck a new posture. Spreading arched and
violently quivering wings, she bowed forward slightly to the cock
(once more on his boulder) with the food-soliciting antic of a juvenile.
A minute and a half's soliciting, however, elicited no response from
the cock, and she flew out to plop into the river, where the two fought
again before going their separate ways.

By the middle of March some dippers on the hill streams were
building nests or had already laid eggs, and the Blacksmith of the
Shingle — as the Gaels dubbed the dipper — was speeding up and
down the Tromie with his mate, carrying beakfuls of moss to a
chamber under a clump of heather overhanging the wall of cliff on the
east side of the rough-stone arch of the bridge, and a foot or two
above the highest flood mark. I never witnessed an actual mating by
any of the Spey dippers, though on an April evening on a hill stream I
had watched two do so when almost submerged on a stone at the
water's edge; and then all that could be seen of them in the fading light
were two brilliant white discs bobbing against the darkly-flowing
water. Yet at mid-March there were still eight dippers occupying
stations on the Spey, though I could not of course ascertain whether
they were present throughout the twenty-four hours; but it seemed
improbable that one or other of a pair would visit their hill-stream
nesting territory for an hour or two's nest-building or egg-laying and
then return to the Spey.

The dippers' song season reached a peak in March, and on the
afternoon of the 20th one was in almost continuous song (with
mandibles slightly parted and the white membrane of its false eyelid
nictating incessantly) for three-quarters of an hour. Another bird
rearing up in song to an alighting mate, and thereby showing off its
white throat and breast, turned upstretched head tautly from side to
side, while rocking slightly in stiff legs. Seen thus, with tail spread, it
appeared three times the size of the presumed hen crouching on an
adjacent boulder. Several displaying birds were in song, and on my
deliberately driving one songster up to another seventy-five yards
distant, an aggressive close-quarter chase ensued. But pursuit-flights

by twos and threes were also as frequent as previously, and one bird embarked upon a high circular flight *alone*, though other dippers were at their stations below. When one of these flew at another on a boulder fifteen yards from it, the latter dived under water before flying two hundred yards upstream. A third bird then flew low over the assailant, which merely crouched threateningly in that bulldog-like posture sometimes assumed by a blue tit. When the assailant subsequently flew fifty yards upstream it, in turn, was chased by a fourth dipper, though it eventually returned to plop down in its own reach.

Although there had never been more than eight dippers present in the control reach since the middle of March, and the majority of these had been solitary birds, there was again intense activity on the 27th when a pursuit-flight was for the first time conducted at a considerable height — forty or fifty feet above the river. At the outset the two participants were twelve feet apart, but they closed as they twirled down to river level, and one tweaked the other's tail when they soared and stooped again. At their coming a third dipper exploded into song from its tree-root station: whereupon the pursuer broke off its chase, to attend to this bird, but was instead chased by it. Another tail-tweaking flight ensued, with the pursuing bird still in song; and when they returned to alight near the tree-root station the owner, still singing, assumed the upright threat-posture, while the other bobbed on a boulder. Then the tree bird struck a new attitude, in which it lifted each leg in turn, shifting its weight from one to the other while remaining in the same place. At the same time it turned its head from side to side and puffed out its white breast-feathers, again like a pouter-pigeon. After a brief period of this rudimentary dancing, it concluded by flicking its wings in a deliberate manner a number of times; nor did it pursue the other bird when the latter ultimately took off from its boulder, but flew with quivering wings to alight in the water, then mounted a stone to flick its wings a few times, and finally began fishing — though this was an activity that had waned during March.

Later that afternoon another 'pair', one of which was in song, also took part in a pursuit-flight of some three hundred yards at a considerable height above the river. After alighting on adjacent boulders one launched itself at the other and they took off on another chase. It was probably one of this pair which subsequently came speeding back, but left the river at a bend and cut high across the grazings towards the hills. This was the first time I had seen a dipper do this, and it was perhaps making for its nesting stream.

Finally, I found that a pair were again present at the tree-root station. One of them was singing vigorously with that particularly loud and sprightly song more commonly heard from a bird on the wing, when the song would be introduced by a burst of rich, throaty *tew* notes and sounded much louder than normal. Indeed, on bending down on various occasions to listen to birds singing just below the river bank, I had always been surprised to hear that their songs were no louder at a range of only two feet than when heard at a distance. As the tree bird uttered his strong and stony yet flute-like notes, while drawn up to his full height in his splendid courtship posture, with snowy breast puffed and pouted, and spread tail depressed, so he turned his taut throat from side to side and intermittently vibrated drooping wings. In the meantime his mate (a paler black and duller white) hopped about from boulder to boulder a few feet away, occasionally feeding; and when she eventually, and inevitably, flew away, she rose to a much greater height than he did, though following a normal course along the river.

There were still two or three dippers on the river in the middle of April, and in one instance when one was about to pass a tree-root station — during the course of a three-hundred yard flight — it, on sighting the occupant, plopped into the water beside it. Thereupon the latter immediately left the tree and flew a further three hundred yards upstream to take up a new station. However, after a heavy fall of snow and a sub-zero temperature during the night of the 24th, no fewer than seven dippers were back in the reach the following afternoon. Were they late nesters, reacting to the sudden cold conditions, or the mates of sitting birds? That same afternoon, a mile farther upriver, I watched a dipper feeding a fledgling, which could not have been hatched later than 6 April from an egg laid about 21 March.

Two dippers were present on the 30th, but by 3 May all the Spey stations had finally been vacated. When would they be occupied again? I had to wait fifteen weeks for the answer to that question. But during their absence I was astonished to observe, on 29 July, on a tributary nesting stream a mile above the Spey, two behaving exactly like those I had watched on the Spey in the early spring. After repeatedly flying after one another from boulder to boulder *both* would finally alight in the upstretched, tail-spread, aggressive posture and begin bobbing, while turning their heads first one way and then the other; and their state of tension was emphasised by the tautness of their white throats and the incessant nictating of their white eyelids. Even when feeding only twelve inches apart they were still aggressive, for one would run down a rock at the other, with back humped and head bowed and threatening. There could be no possible doubt that they were a mated pair, and the fact that both assumed the threat-posture threw some light on the more puzzling behavioural acts of the Spey birds.

Young dippers were as shy as the adults, and I had only to wade past a nest in a dripping peat-bank, three or four feet above a hill stream in spate, for the nestlings to 'explode'. Dropping down into the swiftly flowing water, they at first wing-threshed over the flood, before slipping under and swimming submerged, to come safely ashore on a shingle beach and clamber weakly over the stones. Their immersion was sufficient to darken their most undipperlike, pale grey juvenile plumage to the typical shade of the adult. Although the hen had been collecting food a hundred yards away from the nest, she had apparently observed the 'explosion'; for it incited her to an outburst of song and display to the cock, and then to chasing away all the sandpipers and pied wagtails in the vicinity. The three fledglings had come ashore some scores of yards apart; nevertheless, she fed them without any delay on flies collected at random from beach, bank and water.

Within three weeks of exploding, one fledgling had already moved three-quarters of a mile downstream, and was independent of its parents. As soon as the earliest fledglings were strong on the wing there was a general dispersal from the nest-sites in the glens and on the hill streams; but in view of the absence of dippers on the Spey throughout the early summer, the dispersal did not apparently take them downstream. Indeed, in June and July an occasional adult, accompanied perhaps by a juvenile, would appear at the sources of hill streams — mere runnels in the peat or sward — at altitudes of 3,000 and even 4,000 feet on the Cairngorms.

But to return to the Spey. It was on 15 August that stations were first occupied after the summer recess by one adult in song and by two solitary slate-grey juveniles dredging in the shallows. There was only a tinge of brown on the heads of the latter, and their breasts were a discoloured off-white. Fourteen days later a probable pair of adults, both fishing and in song, were keeping station from thirty to forty yards apart. Although mid-August was reputedly the central period of the moult, both were in perfect plumage, as was a third adult fishing further upstream; and they contrasted with a very immature juvenile, which bore no trace of chestnut colouring on its lower breast. A general reoccupation of wintering stations did not, however, begin until 11 October, when the Spey was in spate and snow was lying well down on the hills. On that day nine brilliantly plumaged adults were present, together with a surprisingly immature juvenile with only a suspicion of chestnut on the belly; but, if immature in plumage, it was fully adult in behaviour, chasing away one of the adults in the familiar lifting pursuit-flight. For the rest, the pattern of behaviour was as in the winter, with the incessant interchanging of boulder stations, associated with pursuit-flights, predominating. On one dipper flying downstream to the first of three occupied stations on a twenty-five-yard length of the reach there was, after the customary leisurely approach, a scuffle and chase that resulted in the station being occupied during the owner's absence by a fifth bird. Another dipper, after making a flight of fifty yards to fish in an attractively 'oily' midstream flood, subsequently took up station with two others, so that all three were within a space of fifteen feet and two of them within five feet of each other. This close contact stimulated one of the three to burst into song, while shivering drooping wings and spreading tail-feathers, though it interrupted its song now and again when flicking its wings. There had, incidentally, been no song between 30 August and 11 October, when on the latter date one bird had, after a chase, alighted on a boulder in the ousel posture and erupted into splendid song in response to the other flicking its wings.

Was it possible to make any sense of the complex activities of this winter concentration of dippers on the Spey? Since the earliest arrivals at the winter stations in mid-August included some juveniles, it was virtually certain that all were local birds from nearby hill streams, attracted by profitable fishing waters. The daily fluctuation in the number of occupied stations, and the tendency for very few to be occupied when the Spey was in turbulent spate, suggested that this was the case, though I could not explain why some dippers should remain on the hill streams throughout the winter but others take up territories on the Spey. Possibly the former did not hold sufficient food during the winter months to serve the full complement of adults and juveniles. Throughout the winter the Spey dippers were intensely sex-conscious rather than territorially jealous because that was the pairing season. Moreover female dippers, like female robins, evinced such masculine traits as song, display and territorial competitiveness. That eccentricity was sufficient to account for the constant and universal aggressiveness, which was so pronounced that a dipper actually opted to vacate its boulder station rather than share it with another, even though the latter was its mate; but, though vacating its own station, it could still remain within the limits of a suitable fishing reach by itself taking temporary possession of a third dipper's station.

16 Wagtails on the Spey

It was in the middle or latter half of March, when the gleaming butter-yellow discs of coltsfoot were first exposed to the sun, that the cock grey wagtails returned to Speyside. Though fewer in numbers than the homing pied wagtails – rarely more than three or four together, nine or ten perhaps along a mile of river – they were no less welcome, for theirs was the first exotic promise of summer to light up this northern strath, when the jet-black throat-square, pale saffron underparts and deep sulphur-yellow rump of a cock, dipping among the dull pink catkins of an alder overhanging the river, were mirrored in the limpid waters of a dark brown pool.

The cocks associated amicably for the most part, catching insects on the rocks and river mud, or performing the extraordinary feat of mounting twelve feet perpendicularly (aligned on a vertical axis from beak point to long tail's tip) to capture a passing insect. Occasionally, however, one would alight on a boulder with a thin little song and stretch up its head aggressively or, with a burst of liquid notes, chase another, before looping away gracefully downriver.

Grey wagtails were not numerous in Badenoch, and I knew of only a score of nesting stations throughout that vast region, with a maximum density of five pairs along an eight mile stretch of Glen Feshie. Whether on the Spey or at an altitude of 1,700 feet at the head of Glen Einich, impounded between the Sgoran and Braeriach in the heart of the Cairngorms, their nests were invariably sited where rocks engineered a waterfall or where a burn or river raced through narrows. On the Spey itself the advantage of this type of location may have been the fact that it was on those reaches that the fly hatched in optimum numbers when the wagtails were feeding nestlings; but it was doubtful whether that was also the case on hill streams.

It was on 18 May one year that I chanced upon the first Spey nest. It had been built in a mossy crack in a slab of rock in the steep green riverbank and a little way back from the stony edge of a brown race of waters rounding a bend. Within call were redstarts, willow warblers, chaffinches and robins singing from the green canopy of birches, alders and rowans. Both parents were collecting numerous beakfuls of mayfly above the fish-jumping race through the rocks, shuttlecocking up in wind-tossed loops and arabesques after higher fly. The grace and beautiful patterns of their aerobatics were matched by the technique that enabled them to snap up on the wing one fly after another, while yet retaining an overflowing beakful. All the fly they required was captured on a 150-yard stretch of the reach, and on returning with their beakfuls both would follow the same circuitous route: first to a tree-stump, then from point to point up the bank and the face of the rock slab, and finally behind the hanging mat of grasses that screened the entrance hole to the nest, which faced north and was in the shade until late evening. Having delivered their loads, they would remove or wait for a shiny white faecal sac. What possible nourishment could the unsubstantial bodies and wings of mayfly provide for young wagtails? Yet, reared on this diet of vitamin D, they would fledge in two weeks.

By the 24th the young of that first brood had left the nest, and both parents were feeding them on the banks and rocks fifty yards downstream, where the glowing saffron cock, skimming out like a swallow over the shallows, would mount to intercept, head-on, fly travelling fast downwind. His grace and agility were much superior to that of a pied wagtail, which was also hawking fly. Four days later the hen was already preparing to nest again. At one o'clock on the 28th, when the cock was feeding the only surviving fledgling (now strong on the wing and dipping in flight like its parents), the hen flew past them but ignored the fledgling. Subsequently, however, she returned to solicit the cock, crouching with wings shivering and tail spiked; but though his crown feathers hackled he did not respond to her advances, continuing to capture fly for the fledgling. Yet twenty minutes later both parents visited a new nest-hole fifty yards downstream from the site of the first nest, and after a few more minutes the cock entered the hole with a wisp of building material. I noted with interest that the hen, after a brief visit to the new nest, flew past the old one without the slightest hesitation.

The first egg of the second clutch was laid on the 31st, and when I inspected the nest on 14 June the hen was sitting tightly on six eggs. Her white eye-stripe betrayed her presence as she sat facing the entrance hole, for oddly enough it was much more conspicuous against her dark bluish-grey head than the cock's. And then on the evening of the 22nd both parents were again hawking for fly with their incomparable dancing flight, and from the opposite bank of the narrow reach I could see the livid red gapes of the nestlings at the entrance hole whenever one of the old birds approached with food.

Since nesting pairs of grey wagtails were so sparsely distributed I was more than surprised the following evening to locate a second pair less than a mile upstream. This pair was also collecting fly for nestlings, with two fledglings of an earlier brood in unpopular attendance; but in contrast to the first pair they were very nervous, refusing for one hour to enter the nest-hole at the sandy edge of a steep fall of bank above another narrows, where a perpetual rise and fall of innumerable fly swarmed low above the glittering, oily-green race of streaming waters. Pairs of sandpipers, gleaming white in the still strong westering sun, shot up and down the river or high up the hillside to their nests among the trees where, in that quiet and peaceful place of gnarled birches, cuckoos called, pheasants 'jarred' and pigeons cooed, while the explosive, stag-like grunts of a sitting great tit caused me to withdraw my fingers hastily from a hole in a birch knot.

By the afternoon of the 27th the four nestlings of pair number one were almost bulging out of the nest, balancing on its edge with difficulty when turning round to defecate; and one had indeed fallen on to the stones below. Both parents were still removing the faecal sacs, and the cock would glide down with one to the water like a gaily painted paper-dart. Since it was a windy day there were no fly on the water, and the old birds were having to collect their insects among the rocks and under the trees. In these conditions they were bringing only small beakfuls to the nestlings, and in fifty minutes the cock brought twelve loads to the nest and the hen ten loads; but the latter also carried an additional eight loads to the fallen nestling. Actually the first four loads she carried were to the latter, and she never fed it and the nestlings at the same visit. She showed unusual intelligence in carrying food to both nestlings and 'fledgling'; but wagtails appeared to be more intelligent than most small passerines. Although I was, for instance, on the road in the Landrover almost every day the very commonly distributed pied wagtail was the one species for which it was never necessary to slow down; nor had I ever found one that had been hit by a car. Even the juveniles appeared to possess this traffic sense.

On the following afternoon, when it was windy and cold, the nestlings were hopping down from the nest-hole to a mossy ledge six inches below the nest to meet the cock when he arrived with food, and then scrambling up again over each other's backs with the aid of frantically whirring wings. In a twenty-five-minute period the cock brought four loads and the hen seven, including four to the 'fledgling' which was hopping about on the bank and rocks. By the next afternoon all had left the nest and were fifty yards downstream, flying strongly and perching in trees; but only two survived the first twenty-four hours, together with the fledgling of the first brood. The following year what were presumably the same cock and hen apparently returned together from their winter quarters, for both were present in the previous year's nesting territory as early as 15 March. At this stage they were shy, disappearing into the trees when I approached. It must be added that some pairs would arrive at a farm ditch in April and remain there for two or three weeks before moving on, presumably to look for suitable nesting sites.

Juvenile grey wagtails began to disperse from their nest localities as early as the middle of July, when they habitually appeared at farms and townships. A pair of juveniles would, for example, arrive at a farm ditch in the middle of August and stay together in its vicinity until the end of September. All the resident adults had, however, left Badenoch by the end of July, though migrating wagtails continued to pass through until early in December.

When walking with my dogs among the low alder scrub on the banks of the river one afternoon two cock pied wagtails approached from different directions, to flight and hover in song above the dogs, when they resembled and also sounded like long-tailed tits. A pied wagtail was always liable to mount up in song when its territory was invaded, and on one occasion, when I was crossing a shingle beach with the dogs, both the hen and the cock of a pair danced and hovered over us in full identical song, for the hens were also stimulated to song by any form of trespass, and their songs were in no way inferior to the cocks'. In June this song-display by both birds was a frequent occurrence when the dogs were in the vicinity of a pair feeding young. The song of a wagtail was mainly a reaction to some external excitement. A cock, whether feeding on the ground, perched on a roof-ridge, or on the wing, would almost invariably burst into a cluster of excited song-notes at the approach of man or dog, or that of his mate or rival cock, and particularly at the passing of a sparrowhawk or kestrel, or even a jackdaw. Both cock and hen displayed a special antipathy towards hawks, though curiously enough cuckoos did not apparently provoke their hostility. A cock wagtail, or a pair, circling high into the air to confront and pursue a hawk was a characteristic sight. Nor were these pursuits a feature of the breeding season only, for I had watched a pair harrying a sparrowhawk as late as the middle of October, at the end of their final resurgence of song, which began on sunny mornings at the end of September.

Pied wagtails appeared to be more inquisitive than most passerines about other birds and beasts, and also inanimate objects, they encountered. A cock would return day after day to peck at its reflection in the wing-mirror of a car, and I recalled an occasion when a party of wagtails were tripping about a bowling green catching insects, and a weasel had suddenly popped up from the gutter, darted out at the wagtails, and raced back to the gutter again. During the next quarter of an hour the weasel repeated this manoeuvre again and again, always dashing back to the gutter after each sally. The wagtails, however, evinced no sign of fear, merely flipping up a few inches whenever the little animal approached them. The weasel appeared to be playing rather than attempting to lure the wagtails by its antics, and the latter were apparently aware that that was the case. Weasels were, incidentally, so rare in Badenoch that I saw only three throughout our years there.

On a spring morning beside the Spey there might be twenty or thirty splendid, sooty cock wagtails homing northwards downriver. At these northern limits of their breeding range they were perhaps most perfectly plumaged, with intensely black mantles and gorgets contrasting with brilliant white masks and snow-white flanks and underparts. There was, however, considerable individual variation: just as some hens were so black and 'smoky' that they could not definitely be distinguished from duller plumaged cocks, in the absence of the latter.

On a mossy bank under the birches three of the cocks were conducting a ceremony, strutting stiffly with beaks pointing upwards slightly, and every now and again leaping a couple of feet into the air. The antics of these small parties of cock wagtails, homing along the Spey, were a feature of March and April, with first one, then two, then five mounting with very loud song (surprisingly mellow at close range) in their dancing, half-hovering flight, or chasing and fighting over the river with a musical cacophony of songs. Their northward passage through the strath would continue until late in April, when dozens might be hawking fly off the river, and as many as sixty counted in the course of an hour's stroll.

It was fascinating to observe the return of an old friend. On 11 March I woke at seven o'clock to find that the previous year's familiar cock wagtail had arrived in the garden at Drumguish, where he had not been the evening before. Within half an hour he was singing from his accustomed station on the roof-ridge, and continued to sing with hardly a second's break throughout the day, for he would sing more continuously during the two or three weeks preceding the arrival of a hen than at any other time. It was nineteen weeks since he had emigrated in the autumn. Where had he wintered? In the south of England perhaps, for an adult ringed while wintering in Hampshire had been recovered three years later in Glen Feshie. At nine o'clock on the morning of 27 March two wagtails were tripping about the newly ploughed field in front of the house. On the previous day the cock had been alone in the garden. During the night, or in the early hours of the morning, his mate had returned; and without any observed greeting they resumed life together — assuming that they had been paired the previous year. They were not a demonstrative pair, and I recorded only two incidents. On a day in the middle of April I noticed the cock passing his mate at some distance in a curiously strained attitude with his head tilted stiffly and a little on one side, and his beak half open. Again, on a June evening the cock, head bowed, advanced towards the hen in a series of short, scraping runs while spreading and closing his white-edged tail; then he flew up and circled over the bank below her, at the same time bending his spread tail first to one side and then to the other.

If, as sometimes happened, another cock visited the garden, our resident, uttering liquid call-notes, would spend hours chasing the intruder around and about from one perching place to another, though without any very great show of pugnacity. As in the case of other species of wagtails this pursuit-flight followed a formalised dipping and looping figure-of-eight pattern.

Pied wagtails were very numerous summer residents in Badenoch. I located, for example, a minimum of thirty-six nesting pairs over an area of some 55,000 acres in and adjacent to Glen Feshie and Glen Tromie, three-quarters of which was uninhabited hill and moor. They nested up to 1,500 feet, and almost without exception in the vicinity of man or his works, whether these were farm steadings, shepherds' cottages and sheep-fanks, shooting lodges, boat-houses, old saw-mills, abandoned forestry camps, bridges, dams, quarries or, as in the case of the highest nesting pair in the Forest of Gaick, the ruins of old townships and shielings, deserted for more than a hundred years; and even beyond these into uninhabited glens and through such uninviting passes as that of the Drumochter. The remotest nesting localities in the high glen were colonised by the cocks within a day or two of the initial appearance of wagtails in the strath during the first or second weeks of March, though it was usually two or three weeks later before the hens joined them. In 1961, after the Glenmore ski road had been opened, a cock arrived at the carpark at a height of more than 2,000 feet on Cairngorm — surely the highest altitude at which a pied wagtail had ever been recorded in Britain.

After they had fledged, some juveniles began flocking as early as the last week in June, though the majority of both young and adults did so towards the end of August. However, there did not appear to be any particular rules governing their departure. Our resident cock might leave the garden before the end of August, though his mate and their fledglings might still be present in the nesting territory in the first week of October, when the hen would sing repeatedly from the roof-ridge. In other years the young might leave before the end of September, while both parents stayed until the middle of October; or the cock might leave then, but his mate linger on for a further fortnight. There were years, not necessarily of open weather, when pied wagtails frequented the Spey throughout the winter and, as we have seen, joined the dippers in picking small insects off the ice of the frozen river. The majority were solitary cocks and were presumably immigrants or on passage, rather than local residents.

17 The Return of the Greenshank and other Waders

IT WAS a paradox of the mountainous Grampians that vast areas of moor and hill would have been inhabited during the summer months only by such birds as grouse, ptarmigan, pipits and wheatears had it not been for the influx of large numbers of summer-resident wading birds from their wintering places on coasts, mud-flats, salt-marshes and more southerly pastures and arable land. No fewer than twelve species of waders returned regularly every spring to colonise the length and breadth of Badenoch from the marshes and shingle beaches of the Spey, and from the glens, moors and lochs to an altitude of over 3,000 feet on the high tops of the Cairngorms, the Drumochter hills and the Monadhliath. But not a single individual wintered, though very occasionally one might flush an immigrant woodcock during the empty months from late November to late January; but though woodcock could be found nesting in every pine forest, birch wood and alder grove up to at least 1,500 feet, all the breeding stock emigrated in the late summer.

It was usually only in the early morning and late evening that breeding woodcock were in evidence, though on one occasion two dropped into the bracken almost at my feet with a sudden fanning of showy chestnut tails — and disappeared! A minute passed before I focused on a large round black eye, and was then able to detect the broad transverse black bars across the conical head, so perfectly did the dark chestnut back blend with the russet bracken fronds and the dark heather tump on which one of them was squatting motionless. It was towards sunset that one would come 'roding', with slow-beating owl-like flight, in and out of the alders beside the river or on a triangular course along a pinewood ride, and I heard again that curious squeaky whistle and soft batrachian croak. Roding might continue into the midnight dusk of June, long after the young had flown from some marshy, tussocky place under the alders. There, at the edge of a dark pool, I had flushed in May a parent that had been brooding her most conspicuously coloured red-brown and deep cream young, which tottered unsteadily through the tussocks, two by two, balancing with the aid of spread blue-quilled wing-arms, bare of feathers — most helpless of all young waders. Rising with a harsh alarm-call, in response to their high-pitched piping, she towered vertically with legs dangling and tail depressed beneath her abdomen, and circled swiftly through the trees, before finally alighting on the river and swimming along below the bushes. All parent woodcock rose in this curious manner when disturbed, as if carrying a young one between their thighs. This was sometimes the case, and in such instances the chick's bright blue shanks could be seen hanging down. Pitching on the ground, the parent would shuffle along brokenly with the tips of her outstretched wings scraping the earth, as she lurched first to one side and then the other; and with a loud, harsh screaming — not unlike that of a trapped rabbit — lure my dogs to follow her away from the chicks, and allow them almost to overtake her, before she would fly a little farther and repeat her injury-feigning.

In a normal year the earliest golden plover, lapwings, ringed plover, snipe, curlew and oystercatchers returned to their Badenoch breeding grounds in February, though one spring, when there was an early thaw, a pair of ringed plover, which were to be expected between the end of February and the middle of March, made a brief appearance on the phenomenally early date of 22 January; but no more were seen until 23 February. If, however, the winter was prolonged, no waders would return before March, as was notably the case that spring with the two-month freeze-up, when not a single individual of any species of wading bird returned until 15 March, which was the day before the thaw began. Redshank, greenshank and woodcock usually returned in March; common sandpipers, dunlin and dotterel in May. Pairs of dotterel and golden plover might indeed return to the Cairngorms when nine-tenths of the ground above 3,000 feet was snowbound and the remaining tenth waterlogged.

The summer-resident waders could also be grouped according to their altitudinal range. Redshank nested only, and sparsely, on the Spey levels below 800 feet. Morning, noon and evening one or more cock redshank would joy-flight for hundreds of yards over the village and links at Newtonmore, and on fine evenings one heard the interminable, snipe-like purring of redshanks mating, while other pairs chased, minute after minute, running swiftly over the bog beside the river and flipping over the drains.

The ringed plovers' range extended a little higher than the redshanks'. Although the earliest ringed plover to return were usually solitary birds, they were very soon in pairs and feeding in flocks of up to twenty on the fairways of the links alongside the river. These appeared to be their exclusive feeding grounds, though I was never able to ascertain upon what they fed. The great majority of the nesting pairs — and there were not very many in Badenoch — were concentrated on the shingle beaches of the Spey, though odd pairs were to be found on the sandy shores of the larger lochs, and just above 1,000 feet in three glens known to me.

Woodcock, snipe, oystercatchers, greenshank, lapwings and curlew normally nested no higher than 1,500 feet, though on the densely herbaged hills east of the Drumochter a pair of curlew might do so at 2,500 feet. On those hills too I saw a single cock lapwing on a large snowfield well above 2,500 feet one May day, and lapwings had indeed been reported nesting around 3,000 feet on the Cairngorms.

Sandpipers and golden plover ranged from under 1,000 feet to 3,000 feet, but dotterel were to be found only above 3,000 feet on the high tops of the Cairngorms, though on the Drumochter hills an occasional pair nested below the highest curlew. These low-nesting dotterel, however, foraged and courted on the bare tops above. To one accustomed to dunlin nesting on coastal salt-marshes it had always seemed fantastic that a few scattered pairs should return year after year, century after century, to the vast waste of the high Grampians. To the best of my knowledge no dunlin nested below 2,500 feet for a stretch of more than forty miles from the extensive heather moors of Dava in the north to Ben Alder in the south. And once they had returned to their territories in the latter half of May, there, like the dotterel, they remained until they emigrated at the end of July. They never came down to the glens or strath to feed, and all we saw of their kind on the low ground were the occasional pair or trip of half a dozen feeding on the banks of the Spey, or at a moorland pool, while on passage in the spring and perhaps again in late summer.

Of the twelve species of waders nesting regularly in Badenoch there were two that I had been particularly anxious to observe in their breeding haunts — dotterel and greenshank; and it was in the last days of March, when I was in the Forest of Gaick, that I heard the familiar *tew-tew-tewk* and saw a pair of greenshanks rise from the edge of the stream, though they quickly alighted again. That same day a cock arrived in the Dell of Bailleguish, midway between Drumguish and Glen Feshie. Restless without a mate, he wandered far and wide. One evening he would be at a loch above the Dell, but mounted to a height on my approaching and made away to Glen Tromie. The following afternoon he would be at a little marsh on the moors on the other side of the Dell. On another day he alighted with a throaty *cloochkey* at the dragonfly pool on the crest of the hill above Drumguish, only to tower away with his strong, flipping flight and pitch on the moor a mile distant.

Never was there a more elusive bird. There were mornings when, from seven o'clock onwards, I would hear him reiterating almost continuously that soft *chuvee, chuvee, chuvee* in sixty-second bursts; then a second or two's intermission would be followed by another tantalising burst, now close at hand, now far off, as he joy-flighted over the moors. But a whole morning would pass without my being able to locate him. On other mornings I would be more fortunate, for when I was in the pinewoods at seven o'clock on 18 April there reached me faintly that persistent *chuvee*ing. A quarter of an hour later the 'song' seemed to come from over my house, and there, sure enough, at a height of some 300 feet, was a greenshank joy-flighting in the grand manner on the periphery of a circle with a radius of half a mile. While flighting he would tower like a redshank, but very much more steeply and swiftly, and hover for a longer period on sharply depressed wings at the summit of his climb; then, more spectacularly, roll over on to his back, like a lapwing, so precipitately that I could hardly follow the movement, before corkscrewing down with half-closed wings and flattening out with lightning speed when just above the moor, only to sweep up again with glorious ease. But after five minutes of this splendid joy-flight he flew away to the brow of the moor again, where, at long last, he was joined by his mate. In the evening, however, he came to the pool alone as usual, for even after her belated arrival the two were seldom seen together. Morning and evening in the middle of May the cock might be seen returning homewards the three miles from the Spey marshes, mounting and falling all the way without a pause, while intermittently *chuvee*ing, and sometimes at such a height as to be nearly invisible to the naked eye.

When, very soon after her return, the hen set about the business of nesting, the two became extremely secretive. Indeed, in the course of a casual study of them the first year I saw them only once during the first three weeks of May; but on 5 May the following year I found one bird present on a Bailleguish peat-hag, which I was convinced was the hen's nesting territory, since it was in that hag that I had seen a fledgling the previous year. Yet though she *tewk*ed monotonously for long periods and occasionally *chuvee*d and though I searched the hag with my two dogs from end to end, I could not find any trace of a nest. After that, I lost touch with the pair for two weeks. On the 24th, however, when in the Forest of Gaick, I had seen another pair joy-flighting together over an elliptical course, rising and falling, sometimes banking instantaneously like swifts, while whistling — one in a high key, the other in a low key — their rich-toned *cloochkey*, with its husky yodelling counterpoint; and when I returned home in the evening the shepherd surprised me by saying that he had found the Bailleguish nest that morning about half a mile from the peat-hag.

So, at noon the next day I was lying on a hump of the moor, a few feet above the grassy dell and only twenty yards from an ash-grey and white greenshank, who was sitting in the shelter of a boulder a score or two of yards from a burn. The whitish-grey granite was the exact shade of the beautiful gunmetal-grey hatching and light and dark herring-boning of her plumage, and of her snipe-like, steep-browed head, with its long and straight, though slightly tip-tilted beak. She sat very still, watching me with her dark eye, whose false eyelid nictated downwards every second without, however, quite covering the eye within its white ring.

After I had lain watching her for some time, I crawled up to within nine yards of her, when she finally rose with a *creek-creek* and circled at a height in wide rounds, *tewk-tewk*ing continuously, though twice joy-flighting briefly and *chuvee*ing. For ten minutes she clipped around with sharp, erratic wing-beats, before being mobbed by a curlew and pitching down on a flat on the far side of the burn, silent. Rising again at my approach, she made off down the glen, but returned when I moved away, and dropped like a plummet to her round of moss, heather and burnt bearberry, which held four elongated eggs, whose stone-olive ground colour was heavily blotched, stippled and clouded with dark brown and pale purple.

At eight o'clock the next morning she was sitting facing the north wind. Not wishing to disturb her, I watched for only forty minutes without putting her up; nor did I visit her the next day, for it was the hatching of the chicks that I wished to observe; but on the morning of the 28th, when one greenshank was *tew-tew-tewk*ing and *chuvee*ing intermittently, I found to my deep chagrin that the nest had been robbed and that not a fragment of eggshell remained. On 10 June the previous year, however, both greenshanks had been on the peat-hag and, when I approached with my dogs, the two had flown round and round with their characteristic clipping flight, while uttering a staccato *tük-tük-tük* and from time to time the deeper-toned *tew-tewk*. When I drew closer to a 'flow' they began stooping at the dogs (and also at me), and the spaniel soon pointed a young bird squatting in the herbage. Picking it up, I found it to be marked with dark brown and greyish-fawn stripes, and to have a black bill an inch in length, and long, strong, seaweed-yellow shanks. When I put it down again, it ran off, waving its two-inch-long pinions, while its parents *tük-tük*ed continuously.

18 Sandpipers and Young Curlew

THE NUMBERS of common sandpipers in Badenoch cannot have been exceeded anywhere in Britain. There were more than twenty pairs on a half-mile reach of the Spey at Newtonmore, and every feeder hillstream was colonised up to its source. Several pairs nested on the shores of Loch Einich, while at an altitude of almost 3,000 feet on the high tops above the loch a pair could be watched mating in May on a boulder in Loch nan Cnapan, upon whose waters miniature icebergs had floated a few weeks earlier.

The Spey itself, and those of its tributaries that were wooded, were preferred to treeless moorland streams, though the marshy bottom of an open glen, such as Gaick, was also heavily populated. On the Spey every boulder seemed to be occupied by a sandpiper, its pearly breast silvery in the sunlight as it bobbed with wings lifted in little frenzies to a hen, bowed forward on her breast; for the sandpipers mated on the first day of their nocturnal or early morning homing, with the cocks hovering, as if they were hummingbirds, before dropping on to the hens' backs. At the end of May there were always numbers of them skimming out over the river's dark peaty pools and then circling back with their curiously hesitant flight to the alder-lined bank, and their lisping and reeling strings of needle-sharp notes were to be heard on all sides. Other piping birds, beside themselves with excitement, threaded their way with arched wings in and out of the weed on mudbanks deposited by spates. There were cocks chasing hens over the shingle beaches with winnowing wings or with one wing elevated like a sail, to the accompaniment of that incessant *psee-ee* with its shrill counterpointing. My memory, searching for a remembered complement to that perpetual ticking piping, finally registered: the penny-whistle piping of a tystie spinning around on a Shetland voe.

Day after day the sandpipers afforded me endless amusement — as no doubt they did every fisherman waiting for trout to rise on northern streams — with their ceaseless circling flights, and also their scrapping; for the cocks fought furiously for minutes at a time, dancing up and down with one or both wings elevated, as each strove to jump on the other's back, both in the water and out of it. But the dominant feature of those spring days was their display-flighting. One, two or three birds might flight — with one setting another flighting — round and round and round in small circles for several minutes at heights of between twenty and sixty feet, banking erratically at sharp angles, now this way now that, with wings quivering or beating alternately, to the inevitable refrain of that repetitive reeling. It was the redshank's joy-flight protracted, with that same erratic action, though the sandpipers' wings were not decurved but held straight or slightly elevated.

Most sandpipers nested in the heather or among the trees a little way back from the river. After the polished creamy-red eggs had been laid in neat rounded nests of moss the hens were quite fearless, bobbing about within touching distance of the intruder, while piping shrilly with wide-open beaks for a quarter of an hour at a time; or luring a dog from a nest at the base of an alder with a dramatic display of threshing wings and rounded tail spread in a full arc of vivid white tips. The chicks left the nest very shortly after hatching and went to cover under bushes or tree roots. So diminutive and inconspicuous were they that it was almost impossible to find them without the aid of a dog. In the hand they were wild beyond belief, struggling and screaming: in marked contrast to young snipe sitting quietly in deep watery herbage and almost buried in the mud. Yet the adult sandpiper was not in the least shy, while the adult snipe was as wild as any bird could be.

Far out on the Spey levels two curlew were flying in great circles, shrieking and whistling, while from the jungle of rush and sedge at the edge of the bog came a persistent musical piping: a slow, plaintive *pee-pee* and *pee-pee-pee*. Not for a moment was there a break in that insistent ventriloquial piping, and it seemed deceptively close as I homed on it for a hundred yards or more through the dense marsh undergrowth. It persisted indeed until I was right on top of the piper, which was stalking long-legged through the reeds and tussocks in a lost sort of way, and gently captured that delightful creature, a young curlew, two or three days out of the egg. When I picked it up it changed its plaint, emitted with short beak closed, to a hoarse, almost inaudible rendering of its parents' *courr-lee*.

That soft and melodious, yet far-carrying piping of young curlew was one of the lovely sounds of June in the Highlands, though the loveliest was the slow, clear *courr-lee, courr-lee* of their parents. But, when danger threatened, this changed to an anxious *quew-quew-quew-quew*, as they circled wildly over their running young and zoomed down low over my head with wings half-closed, though never pressing home their attacks as boldly as the lapwings. One parent might even perch, distraught, on the top of a tree.

No matter what else I might be doing at the time, I was always listening for those curlew cries from mid-May on, to tell me that the young had hatched once again, and that it was time to prepare for ringing. However, young curlew were not easy to find in any numbers in Badenoch, for the loosely grouped colonies of two or three pairs of adults were widely scattered, and the chicks would run very fast in different directions, before finally freezing against the exposed roots of a tree or under a fallen log. Moreover, when the latter were only a few days old, their parents would shepherd them down from the moors or up from the marshes into the intervening birchwoods or groves of aspens, though they themselves would still circle noisily far out over marsh or moor, and cunningly mislead those would-be ringers who were not familiar with this aspect of curlew behaviour. Nevertheless, it would be no exaggeration to say that it seemed a miracle that any young curlew could survive those critical first five or six weeks, so fraught with danger, before they gained the freedom of their wings; for throughout that period they would stalk openly about the mossy dells among the trees, and instead of crouching down at their parents' warning cries, as most young wading birds did, would run awkwardly up the grassy banks. Only when all was lost, and one was upon them, did they flatten in the grass or against the bole of a tree; and then indeed they were well camouflaged in their buff-coloured down with dark brown spots. Oddly enough, the few young curlew on the almost treeless moors of the Drumochter were the easiest to ring. There, they stalked about tamely on either side of the road bisecting the moors, while their parents, accustomed to the endless stream of traffic, did not guard them with anything like the ceaseless vigilance of those in less populous haunts, nor become agitated when I ventured to approach their gangling infants, but merely stood quietly at some distance.

Although the earliest young curlew did not hatch until the middle of May, adults — presumably non-breeders or unsuccessful ones, and many with gaps in their flight feathers — were already passing west up-Spey by the end of the first or second week in June. They would continue to do so throughout July and August and into the latter half of September, with an occasional late bird perhaps in October. At any hour of the day, though especially in the afternoon and evening and as late as eleven o'clock twilight, they streamed westwards in sixes and sevens and flights of as many as nineteen. Most followed the course of the Spey in bunched gaggles or Vs, and with almost every flight there was a laggard hurrying to catch up with its companions. Normally they migrated at a height of some hundreds of feet, but very high in fine weather with a north-easterly breeze, when, with clear skies after a cloudy interlude, their passage would be heavy. In those conditions they flew at such a height that they would have passed unnoticed had it not been for their persistent contact-calling — that nostalgic *phwee-phwee-phwee-phwee* — and an occasional joyful skirling. Whenever in a bad summer, associated with days or weeks together of south-west winds and a persistent pall of cloud on the hills, there was an hour's change to a north-east wind tearing blue rents in the heavy cloud masses, then immediately the clear spaces would be etched with westering flights of curlew. So anxious were they to emigrate that they took advantage of every fair interval, no matter how low the cloud base on the hills. In such weather conditions the Spey was likely to be in flood, and it was noticeable that the curlew, rather than adhering to the serpentine course of the river, might be flying half a mile to one side of it.

19 Lapwings on the Moors and in the Glens

EVERY spring February and March presented a problem that I was never able to solve. One February afternoon, for example, I was surprised to see a flock of one hundred and twenty lapwings come *down* the Spey with their usual direct and silent flight from the *south-west*, and alight in a favourite riverside field, where another fifty were already feeding. In the past I had observed the occasional small flight of lapwings passing through the strath from south-west to north-east, but these had not amounted to more than one per cent of all I had recorded. So, a flock of more than a hundred arriving from the south-west was noteworthy; for a feature of the spring return to nesting grounds in Badenoch, not only of lapwings but also of oystercatchers, curlew, golden plover, ringed plover and redshanks was that virtually all one saw — and these were predominantly lapwings, which migrated mainly by day — came *up* the Spey from the *north-east*. At two o'clock on the last afternoon of March one year, for instance, I was fortunate enough to witness a flock of ten ringed plover flight in from the north-east with a hardly credible jinking and banking to alight on a fairway beside the river. The exuberance of their fly-in suggested: home again! Similarly, in the third or fourth week of March — never earlier than the first week — the earliest returning redshank might be seen rocketing steeply down to the river from the north-east.

This north-east to south-west immigration of waders could be traced back as far north-east as Moy, only a dozen miles from the Moray Firth. Although numbers of lapwings, together with other waders, wintered on the spacious farmlands and airfields of the Moray Firth forty to fifty miles north-east of Newtonmore, one could be as certain as one could of any problem concerning birds that these Moray winterers were not born and bred in Badenoch, if for no other reason than that all recoveries of waders ringed as young birds in Badenoch had come from the far south and west. Thus one of the few

137

young redshanks I ringed in the Newtonmore area was subsequently recovered on the Solway Firth in September the same year, and a young curlew on Islay the following spring, while of nearly 1,500 lapwings ringed mainly in Badenoch, with Newtonmore as the centre, fifteen were subsequently recovered at a distance: one in Portugal, two in northern Spain, three in south-west France, seven in Ireland and two in Ayrshire. Five of them were first-winter birds, one Irish recovery was in its ninth winter and another in its sixth, and of the two Ayrshire recoveries one was in its second winter and the other in its second September and perhaps still *en route* to winter quarters when it collided with a plane at Prestwick. Similarly, of twelve distant recoveries from 540 Badenoch-ringed oystercatchers, six were from Ireland, one from Wales, two from Lancashire and three from south-west Scotland. Nine were recovered during their first winter, two in their second or third winters, and one in its fourth winter. Moreover, there were no records of any Highland-bred oystercatchers wintering anywhere except south and predominantly south-west of their breeding grounds.

The evidence of these ringing recoveries suggested that all, or the great majority of lapwings and oystercatchers, and presumably other waders breeding in Badenoch, wintered to the south and west. Therefore their most obvious route home was in from the west coast by way of Loch Laggan, and possibly by the Drumochter, since lapwings nested in all suitable places throughout that Pass; yet in practice they apparently approached Badenoch from the Moray Firth, which they must have reached by way of the Great Glen and then homed south up their appropriate glens into the Grampians. Although there did not appear to be any records of such a passage up the Great Glen, I had myself witnessed a flock of lapwings travelling from west to east, against an easterly gale, along the edge of a high tide at Lossiemouth on the Moray Firth during a thaw at the end of February. If they had flown a further eight miles east they would have reached the mouth of the Spey.

The immigration of the summer resident lapwings to Badenoch, and the taking up of nesting territories, extended from the middle of February to the first week in April, but in open winters was concentrated mainly in the last week of February. January and February, and sometimes March as well, were the hardest winter months in Badenoch, with even the low ground frostbound or under snow for weeks at a time in a severe winter. Normally, there was no immigration in such conditions. But once the long winter freeze-up had broken in February or March there would not subsequently be any prolonged hard weather, though there might be occasional heavy snowstorms well into April. Those, however, the resident lapwings survived, courting and scrapping noisily on steep braes from which the snow had melted. If snow fell continuously after they had laid their eggs, they would incubate throughout such storms and, when they rose from their nests, left conspicuous snowfree mounds. With snow almost every day during the first half of April one year, culminating in a fall of several inches, I found only one nest temporarily deserted; but despite the fact that there was frozen snow in the nest and that the eggs were stone-cold, the latter nevertheless eventually hatched out. A hard spring was, however, likely to result in a higher than normal incidence of infertile eggs.

The return immigration took place mainly in the morning and early afternoon — exceptionally in the evening — and migrating flocks might include up to one hundred and fifty birds. The majority of the flocks would flicker leisurely up the strath in wavering lines and shifting bunches — though some flew fast and direct — and invariably in absolute silence, in marked contrast to the homing flocks of joyfully piping oystercatchers. Some flocks might pitch down on the river beaches and Spey levels or on adjacent stubbles and bare pastures to scrape at the dung patches; other flocks would pass on westwards upriver. Indeed, of one arriving flock half might drop down to drink and bathe, while the other half continued on their way. Those that alighted might be joined by solitary residents which, however, returned to their nesting territories if the newcomers rose again to continue their migration, though not before they had accompanied them for a short distance. One often witnessed this phenomenon of flock-splitting, and also that of two migrating flocks merging in flight. Clearly the composition of any one flock of migrating lapwings must have changed repeatedly during the course of their travels.

Most flocks passed low along the Spey, which was their usual fly-line. But occasionally a flock would come up the strath at a great height and circle over the village for some minutes before passing on; or a flock of perhaps thirty, which had alighted, would suddenly mount up from the riverside, soar in an easterly direction to a height of several hundred feet, and then peel off westwards, tumbling with extraordinary velocity, before journeying on upriver in two flocks while performing remarkable flight evolutions.

The majority of the first homing lapwings to return were cocks. The earliest of my ringed birds — of which I distinguished upwards of a dozen in subsequent years — to return to their territories during the first three weeks of March were all cocks. The nesting density on the Spey grounds was very high. On my main ringing beat around Newtonmore an annual average of from forty to fifty pairs nested on twenty-five acres of links, sparsely heathered bog and rough grazings beside the river. Twenty miles north in the vast depression of the Carr, where hundreds of acres of bad old pasture lay adjacent to bog, poor arable land and an immense shingle beach, the density of both lapwings and oystercatchers was even higher. By the number of lapwings one could gauge the standard of farming. But only ten miles west of Newtonmore one passed right out of lapwing country, and few more would be seen all the way to the west coast. It was difficult to account for the scarcity of lapwings in the West Highlands, where there was a superabundance of old crofting townships — abandoned for a hundred years and more — and their relics in the form of green 'parks' and sheepfolds, which in Badenoch were one of the lapwings' favourite nesting habitats. Presumably their absence in the West could be attributed to the scarcity of arable cultivation, though it had to be noted that there were many high glens and shielings in Badenoch that lacked arable land but which were nevertheless well populated by lapwings; while here and there a pair or two were to be found nesting in a little green oasis among the heather and peat-hags on what had once been a hill crofting, with only the mountain hares for company.

The lapwings' distribution in Badenoch was almost identical with that of the oystercatchers, except that only the odd pair of lapwings nested on the riverine beaches — the main breeding grounds of the oystercatchers — and, conversely, only the occasional pair of oyster-catchers nested on the rough grazings, stunted heather and bog adjacent to green pastures preferred by the lapwings. Normally lapwings nested no farther up the hill than oystercatchers, nor any farther up the glens, with the altitudinal limit for both lying between 1,200 and 1,500 feet; but as that limit was approached, so the nesting density decreased to a level at which one colony of a dozen pairs of lapwings might be six miles from the next colony.

Some of the earlier homing lapwings proceeded direct to the higher nesting glens. Initially, however, they probably visited these only at night. At Drumguish, for example, I *heard* the first lapwing one year at 8.30 p.m. on a mild, wet night with a dim moon as early as 11 February, and every night thereafter between 7.45 and 10.30, when it usually called once and no more; but it was 10 March before a lapwing arrived at the township before the darkening — but did not call until moonrise at 11.30 p.m. — and noon the following day before I *saw* the first flock of twenty-five at a frozen lochan on the hill above the township. March that year was a very severe month.

At Drumguish I found that the arrival of one or two lapwings at moonrise, to call for a few minutes between 8 and 11 p.m., was a feature of their spring return to the higher glens — as indeed it was of other wading birds. On mild nights in February and March, when we sat with the window wide open to the still night until bedtime — and were, even so, too warm — my subconscious would be alert, all the hours that I was writing, for that first few minutes' calling of lapwings, curlew, snipe and oystercatchers; and then silence until dawn and the next night. But once settled in their territories then, with the moon at the full and weather conditions suitable, the lapwings would call all night long in March and April until dawn. Their wild crying, screaming and *wulloch*ing, together with the urgent thrumming of tumbling wings — now near, now far — echoed from field to field and through the glens hour after hour, varied from time to time by that characteristic night sound of the lapwing — a persistent, querulous squealing. Almost as continuous would be the 'creaking' and drumming of snipe and, more intermittently, the communal piping, *peek*ing and rippling of oystercatchers, and the soft *tirr-phee-ew* and skirling of golden plover and curlew. For those who lived at edge of the moors this was the supreme consolation for the absence of the morning and evening chorus of small song-birds; but when one awoke in the morning sun, all would be quiet except for the rolling songs of chaffinches and the reiterated *kopack-r-rr-rr* of grouse. Even on warm sunny days there was not a sound from the waders, and before the darkening only the whistling of a song thrush and keen 'scythe-wetting' of a partridge, until, in the gathering dusk, a snipe would begin to 'creak' and release the full waders' chorus, which came to an abrupt end with full darkness on nights when there was no moon.

Although, as April drew on, snipe normally drummed only at the darkening or from moonrise on through the night, light showers of rain stimulated all the cocks to joy-flight on hot sunny days; but by mid-May, when the hens were sitting hard in little rounds of dead grasses on their olive-green eggs coarsely stained and mottled with burnt-brown blotchings, drumming was falling off, though there was much 'creaking' at all hours up to the end of June; and even in the last days of July a shower or a few thundery drops would stimulate the cocks to mount and drum briefly.

No doubt the habit of first returning to breeding grounds at night was common to all those lapwings that nested away from the Spey, except possibly for those near the altitudinal limit and ten miles or more from the river. However, even those outposts might be colonised as early as the middle of March, and one year I recorded the first lapwing of the spring in such a glen; but normally there was probably an interval of from nine to seventeen days between the earliest immigration to the Spey grounds and the first diurnal visits to glens quite near the main strath, while three or four weeks might elapse before the highest glens were visited by day. In this respect I observed that the occupants of a newly colonised site in a pinewood that had been clear-felled eleven years earlier, did not return to it diurnally until six weeks after the first birds had arrived in the strath, only two miles away, and a month after the first pairs had taken up residence in one of the higher glens. If a prolonged spell of hard weather intervened, after one of the latter had been colonised by a score of lapwings, the birds were noticeably inactive and silent, and early in the afternoon would flight away in small groups in the direction of the Spey.

In an early nesting year the first eggs were laid about 22 March. They were laid on alternate days, and were incubated for between twenty-five and thirty-two days. The first young hatched at the end of the third week in April and left the nest-scrapes an hour or two after hatching. They might have moved ten yards away from the nest before they had dried off, seventy-five yards after two days, and a hundred yards after four days; and they could swim strongly within twenty-four hours. After a fledging period of about six weeks the first young were on the wing in June, a little before or after the earliest adults to moult were packing in dozens or scores beside the Spey, and when small flocks of adults were already passing south through the high glens, though not through the strath. Perhaps some young ones emigrated with their parents, though a feature of late summer from the middle of July onwards was that of discrete flocks of as many as two hundred young lapwings — my locally ringed ones with well-forked crests among them — which frequented the links and beaches. A possible commentary on the severing of family ties after the breeding season was afforded by the unusual incidents of one autumn when, from the end of August until the middle of September, a family of two adults and four young scrapped incessantly day after day, with the latter persistently attacking both parents; while at the darkening every night there would be more excitement and fighting, indicated by the noisy calling of the adults and the squeaky rippling cries of the young ones. However, by the end of the first week in July almost all the adults would normally have left the strath, though at that date a few chicks would still be hatching from replacement clutches, and odd pairs of adults might linger at lower hill stations until the end of the month, while here and there a solitary cock would stay until late August. Sometimes what was presumably a resident pair would reappear at one of the hill stations *at night* after an absence of several weeks, and on one occasion a cock remained in his territory until the extraordinarily late date of 10 November. Yet throughout June and July not a single lapwing was to be seen leaving or passing through the strath. Their departure must therefore have taken place at night, in contrast to their arrival in the spring; but what constituted night to a lapwing when the summer darkness in the Highlands was no more than a twilight?

With virtually the entire local stock of both adults and young having departed from Badenoch before the end of July, the autumnal passage of lapwings was sporadic and varied from year to year. One autumn might go by without a single flock of migrants. In another year it might be the middle of October, with low cloud rolling up from time to time to disclose patches of snow on the black mountains, and the river in full spate, before the first lapwings seen since the end of August — a flight of four — would come tumbling in from the west at dusk to alight on the links, from which snipe exploded on all sides. In other autumns there might be passage movements on as many as twelve days, mainly in August and September, but continuing until the end of October. Even so, the total number of lapwings recorded would have amounted to no more than six hundred.

In contrast to the spring immigration and passage, which was almost exclusively from north-east to south-west, this autumnal movement might be in either direction; and, with the infrequent exception of solitary adults, apparently consisted of juveniles. Those flights heading north-east appeared to migrate more swiftly, purposefully and directly than those going south-west, and flew so low that they were sometimes lost to sight when rounding a bend in the river. By contrast, a flock of two hundred, wavering south-west against a stiff breeze, might circle back to some inviting field, and only a dozen of them pass on. Similarly, groups from the many large assemblies of juveniles feeding on the fairways after noon would mount up from time to time and carry on westwards.

There was one other aspect of the lapwings' autumnal movement through Badenoch. Long after the nesting grounds had been abandoned, and when no lapwings were to be found anywhere in the strath, I would go up to one of the highest glens and find it full of juveniles only; and in the spring, when the roads were open once more, I would learn from keepers or shepherds that a flock of these immigrants had wintered in the glen for several weeks. Again, around Christmas or after the New Year, when northern Europe and England, but not the Highlands, were experiencing very severe weather conditions, an occasional flock of perhaps sixty lapwings might pass silently west up the Spey. Such hard-weather passage from north-east to south-west might continue into February, though the movements of those taking part appeared uncertain. Coming in high from the north-east a flock of fifteen would beat up and down the river, circling and whiffling for as long as half an hour, before finally alighting or passing on south-west.

20 Oystercatchers on the Spey Beaches

AT THREE-THIRTY on the afternoon of 9 March a flight of thirty-seven oystercatchers passed west up the Spey with a chorus of chattering and an occasional pipe, flying fast and low above the river, though mounting to a height at bends. Half the pack pitched down to alight at the edge of the river, while the remainder continued on their way.

This was a most unusual occurrence, for the oystercatchers normally returned to their nesting grounds in Badenoch at night or, if not at night, then before I was out of bed in the morning; and I did not lie abed in the spring. Over the years indeed I recorded only half a dozen instances of diurnal spring passage. Usually the first ever-welcome indication that our local summer residents had returned once again was the shrill piping of one circling unseen over the house at some time between eight-thirty and ten at night. Such an arrival hour suggested a dusk or after-dusk departure from the Moray Firth — assuming that the Firth had been its previous staging post — though, as we saw in the preceding chapter, it was improbable that any Badenoch oystercatchers wintered there. Where, for that matter, did each batch of young birds spend their second and subsequent summers? Not until nine years after I had begun ringing them did I observe an adult with a ring in the Newtonmore area, and then only for a single day (26 March); and it was the tenth year before one ringed adult returned to nest, as it did the following year too after arriving as early as 11 March. The latter was a cock, whose mate nested only three hundred yards from my house. Over a period of seven years I had ringed sixteen young birds from that nest-site. I could not be certain that the ringed cock was one of them; but from the familiar way in which it came up from the nesting bog to feed on the fairways above, to which successive years' young had always been led by their parents, I had little doubt that it was. One would have supposed that after a lapse of eleven years many more than one or two out of more than five hundred ringed young would have survived to return to their Newtonmore birthplace to mate with some of the established adults, even though oystercatchers were very long-lived birds, not breeding until perhaps their fourth summer, and with records of recovery twenty-seven years after ringing. Among waders only curlew — which had been recovered thirty-two years after ringing — might have a longer life-span.

Arrival by night set the pattern of the oystercatchers' behaviour. From the time of their return until the end of April a feature of their life on the nesting grounds was their exuberant piping on moonlight nights, or at the darkening on calm, mild nights, and again at dawn. At those hours a trio of oystercatchers would indulge in such a frenzy of sustained 'rippling' that anyone not well acquainted with their kind would have supposed that twenty or thirty were displaying, instead of a pair and a third bird hawking around with that slow flapping joy-flight, associated with a continuous *peek-peek* or *kirree-kirree*. However, hard frosts or snowstorms more or less silenced them. If their nocturnal arrival was missed then their first appearance would be at some well-known resting place on their nesting beaches or on the river bank near to these, for in the intervals of feeding on their favourite stubbles or pastures — which might be white with snow — they would spend much of their time in those early days sleeping at the edge of the river. Some of the earliest arrivals were lame birds, which might have wintered nearer home than their fellows. In an open spring, when at least some pasture and arable land was clear of snow and had thawed out, local residents might return as early as 20 February; but when hard-weather conditions prevailed far into March then none might do so until as late as 21 March. In February-returning years the first-comers were likely to be solitary birds or pairs. In March-returning years half a dozen might arrive the first day and a score the next, and maximum numbers would be reached within five to twelve days, with flocks of from fifty to a hundred or more feeding on the stubbles, though there was usually very little ploughing in Badenoch before the spring thaw.

The first birds to return were probably those with territories in or near the strath, though one could not be dogmatic on that point because it was difficult to visit the high glens on their precise arrival date; and it was quite possible that the latter were visited only at night initially, as appeared to be the case with lapwings. It was certainly my experience that no oystercatchers were to be seen in subsidiary glens until from five to twenty-one days after the first had been observed in the strath, though I believed that in March-returning years the lower reaches of these glens might be occupied immediately after their arrival in the strath. In any case these glen birds probably flew down to the extensive water-levels and pastures beside the Spey in the late afternoon, for as late as the middle of April flocks of as many as sixty oystercatchers could be watched flighting down-glen to join the noisy companies of their fellows already feeding on the levels.

Individual nesting territories were first occupied for a part of the day at the end of the third week in March, and permanently so a fortnight later, even if the weather was wintry. The vast majority of the pairs nested on the stony beaches in or alongside the Spey, though hillocks, high ridges and even heather flats, as far as a thousand yards from the nearest stretch of river, were also favoured. Elsewhere in Badenoch, and northwards to the Moray Firth, very many pairs nested in cornfields, often at a considerable distance from water. The nesting density on the Newtonmore beaches was very high, and on one 1,500-yard stretch of the Spey 265 yards of beaches were occupied by an annual average of twenty-five pairs. Nothing approaching this density was to be found anywhere in the glens, where at the altitudinal limit of 1,500 feet a mile of river would hold only two or three pairs, nor on the Moray Firth and its hinterland. On the beaches it was noticeable that the nesting pairs tended to be grouped in twos, a few yards or as little as six feet apart. In one such instance two neighbouring hens each laid clutches of four eggs. They were probably related, for I recorded only six other clutch-fours during the eleven years that I was ringing. In most cases the same territories, or even nest-hollows, were occupied year after year, and one nest was used in twelve successive seasons. Once I had marked down all the territories along a given stretch of river, then I did not find more than one or two new nesting sites the following year; nor more than one or two of the previous year's territories unoccupied.

The first eggs were laid on alternate days during the latter half of April, invariably by birds nesting alone in heather or bog some hundreds of yards from the river beaches. In six years out of eight one isolated heather-nesting bird produced the first clutch on the breeding grounds between 14 and 23 April, while in the other two years she was preceded only by a similarly isolated bird. The earliest young hatched shortly after the middle of May, from twenty-six to thirty-four days (commonly thirty to thirty-one days) after the first egg laid, though thirty-six days might elapse if the parents had been constantly disturbed by passers-by; and a clutch of four might all hatch successfully. The young left the nest within a few hours of hatching, and might have wandered as far as two hundred yards from the nest before they were five days old, and a quarter of a mile before they were able to fly; but the nest-site remained their focal point, and some might still be found in its vicinity after they could fly. As soon as they were strong on the leg the young would scuttle to cover or to the water at the warning piping of their parents. Putting its head down, one might deliberately swim under water for a few yards, rowing itself along with its wing-arms, though it would have considerable difficulty in keeping its stern down. It would swim strongly across a rapid current, though surfacing frequently, while one parent would be beating outstretched wings on the ground, or rocking them from side to side, if it happened to be one of that small minority among oystercatchers which reacted in that way to the presence of a man or dog. A few parents were extremely bold, flying directly at my head and alighting only a few feet away, to rock slowly spreading wings as they squattered along the ground. All parents of course false-brooded incessantly, with a jovial look of complacency in that sparkling crimson eye as the innocent ringer was lured from the crouching chicks; and this they might continue to do after the latter were on the wing.

The earliest young flew at the end of the third week in June, from twenty-six to thirty-one days after hatching, though there were still a few unfledged young about as late as the first week in August. They, presumably, were the produce of second clutches by birds that had lost their first clutches; but very few oystercatchers laid replacement clutches, and I recorded only three definite instances of this. However, by the end of June the Spey beaches would be almost deserted, for the parents led their young on to the bog and rough grazings and into the crops of hay and corn and turnips; and at dusk every evening one well-known parent whose eggs I had found, and most of whose young I had ringed, for eleven consecutive summers, would bring the latter to feed on the links at my garden gate. Her ice-keen *peek-peek* would be a pleasantly familiar night sound for the remainder of their stay. It was possible that the family fed there throughout the night, for I would sometimes see them when I got up in the morning. At midnight, and then again as early as three in the morning, I would hear once more that joyous rippling and piping of oystercatchers display-piping up and down the river. Ironically, this was especially pronounced if they had had a bad season and had lost large numbers of eggs and young. In those circumstances a flock of fifty adults might gather to feed on a fairway — all parental cares over for the year.

I assumed that it was the hens which remained with the young for so long after they fledged, for although some parents continued part-feeding them for weeks after they were on the wing — some would still be doing so in the last days of August — the departure of the adults began in the middle of July. Exceptionally indeed, oystercatchers might be seen passing through as early as the third week in June, and the majority of the local birds had gone before the end of July. In August one came across discrete flocks of ten juveniles, while flocks of twenty might include five or six adults. The juveniles passed much of the midday hours sleeping on a favourite shingle spit, which was also a gathering place for young lapwings and black-headed gulls, and a temporary staging post for migrating oystercatchers. No doubt these gatherings on the beaches and adjacent feeding grounds played a part in stimulating the departure of the local juveniles, most of which left before the middle of August. At eight o'clock on the evening of 20 August one year, for example, six juveniles flew in to alight on a fairway. Twenty minutes later I noticed that their numbers had increased to thirteen and included one of my ringed birds, which displayed briefly with two of the others. But by the next morning all had gone. Thereafter, the majority of oystercatchers remaining in Badenoch were either very young birds or cripples with broken wings or legs. Thus, one young ringed bird with an injured wing, which had fledged in the middle of July, was still with us two months later; while a lame bird was present from late October until early in December.

During the years that I was ringing I watched some thousands of oystercatchers passing up-Spey during the late-summer dispersal from their breeding grounds, but not once did I witness any of the local birds actually take off on departure. Why was that? Let me remind myself once again of the observed facts of this post-nesting dispersal.

It was normally in the middle of July that oystercatchers began to pass west — all invariably west — along the course of the Spey through Newtonmore, and continued to do so until the last days of August, when some flocks were composed exclusively of juveniles. Although the earliest of these westering flights might exceptionally pass through at noon, the majority did so between six and ten in the evening, and more particularly between seven and nine o'clock.

Oystercatchers, when migrating, kept up a continuous noisy piping, and my house lay right on their fly-line; yet I did not record a single flight before noon or after eleven at night, though on a fine night it was still light at midnight.

Fifty per cent of the flights did not include more than fifteen birds, though flocks of up to fifty and, in one exceptional instance, one hundred and twenty, were recorded. Most flew less than fifty feet above the river, but on fine clear evenings might do so at such a height as to be almost out of range of the naked eye, and would have passed unnoticed had it not been for their incessant piping; but even when flying low they always mounted up at river bends, and temporarily leaving the river, cut across the bend to the next straight reach, maintaining their westerly course. They usually flew fast and direct, though frequently a flight, or a section of a flight, would drop down to a beach, whereon residents were always sleeping or bathing, and rest there for half an hour before taking off again, to an encouraging outburst of piping. Again, however, often only a section of the original flock would take off, and the remainder delay their departure for some minutes. The musical chatter of passing flights might incite a parent, tending her young by the river, to answer, and they would respond by circling above, and then circle a second time before alighting on the beach. Ten minutes later they would take off again, with two long indecisive circlings, before being set on course by another noisy flight coming up from the east. As in the case of the lapwings, so the composition of these migrant flocks of oystercatchers must have changed many times while *en route* to their winter quarters, and it was possible that no two individuals completed a migration together.

Their westerly passage was never a regular evening entertainment. Given reasonable weather conditions, it averaged about one evening in three over the whole exodus period, since it was restricted to fine clear evenings or clear intervals, which were usually associated with a shift in the wind into a north-easterly quarter. During spells of cloudy weather several days might elapse without a single flight appearing, though when the cloud lifted after some days of rain or gales as many as eight flights, involving about one hundred and fifty birds, would pass through in one evening at fairly regular twenty-minute intervals, and continue to do so until after twilight. (It was under such conditions that I recorded that large flight of one hundred and twenty.) Oystercatchers were even less likely than curlew to migrate in bad weather. The very hour that the wind backed into the north-east on 27 July one year, for example, after nine days of south-westerlies with persistent cloud, two flights of oystercatchers appeared at 9.40 p.m., although by that date only one local family was still in residence. But the sky had to clear from north or east; with a clear sky to the west, but overcast in the north or east, there would not be any passage.

Eleven years' observation had left me with three major problems. First, why did I never witness the departure of any of the local residents? Second, where did the passage birds originate? And third, why was migration restricted almost entirely to the evening hours, yet never prolonged after dark? To the first problem I had no answer. To the second I could suggest that it was highly improbable that they were all local residents that had collected during the day, and on preceding days unsuitable for migration, on the five thousand acres of the Spey levels extending from Loch Insh to Newtonmore. There was, however, one clue to the origins of at least some of the migrants; for on 31 July one year I had had the possibly unique experience of picking up on the Newtonmore links a young oystercatcher that I had myself ringed the previous 2 June at Llanbryde, a few miles inland from the Moray Firth and fifty miles from Newtonmore. On the other hand, another young bird ringed on 16 July at Rothes, a dozen miles south of Llanbryde, had dispersed (a little north of east) to Buckie, fourteen miles distant, by 20 August.

The recovery of my Llanbryde juvenile at Newtonmore bore upon the third problem. As in the case of the lapwings, so for the oystercatchers a direct flight from the Moray Firth nesting grounds would have taken no more than two or three hours. Yet, as we have already noted, the main migration through Newtonmore took place between seven and nine in the evening, though flights might still be passing until eleven o'clock; but once darkness fell all migration ceased, in contrast to the nocturnal arrival in the spring. Presumably, therefore, those flights that passed through Newtonmore shortly after dusk rested during the night on beaches further up the Spey. I attempted one year to trace their fly-line in a Landrover; but it was a summer of persistent south-west winds and heavy cloud, and I was only able to establish that, as expected, they bypassed the first river leading south to Drumochter, and continued on westwards. But why did they not return by this westerly route in the spring? My quest also indicated that it would not have been as easy to follow their flights as I had supposed, since it was evident that the migrant flocks were constantly dropping down to rest on attractive shingle beaches, and that such rests could be prolonged. On putting up one flock of thirteen that had rested for a long time I was interested to note, once again, that only seven resumed their passage, for the remaining six alighted again after circling around.

21 Golden Plover and Dotterel on the High Tops

WHEN ALL the Grampians were a colourless wasteland; when prolonged frost and desiccating east winds had burnt up pastures and moors alike, and hungry deer and sheep had eaten down the last green blades of sedge; when peaks, ridges and corries were shrouded with smooth gleaming spreads of frozen snow, then the golden plover returned to Badenoch.

On such a March day, when following an old drove road from Lochaber into Badenoch, we came upon a shrivelled yellow oasis of pasture at the moor's edge, sheep-sick and pocked with mole-heaps, derelict with tumbled folds and dykes. The gaping croft-house creaked in the bitter north-east wind, though the granite barn stood firm. It was a sad, long-abandoned relict of man, still protected from the evil eye by its rowan, pine and alder — one of each — and a sycamore with a hoodie's nest. But it was a storm-haven for a pair of song thrushes and a pair of pied wagtails, for wind-blown pipits, small bands of starlings, peewits nesting in the runches and six pairs of golden plover, gathered here to await the coming of spring to the desolation of fire-blackened moor. Its imminence incited them to broken yodellings while making short runs at one another with feathers hackled.

The first of the breeding golden plover might return to Badenoch in mid-February. They arrived both at night and during the daytime, when yodelling flights of up to forty would come up the Spey very fast from the north-east. Some continued on their way; others alighted on the Spey pastures where, by mid-March, flocks of as many as two hundred and fifty would have collected. In late April and early May small flights of the northern race dropped down to rest and made themselves conspicuous by their peculiar habit of *sitting* in flocks on braes and fields. The summer residents were sparsely though widely distributed, averaging perhaps one pair to the square mile, and were probably more numerous around 3,000 feet on the eastern hills of the Drumochter than anywhere on the moors, except for those that had been too closely burned in previous years and were now bare overgrazed sheep-runs that included much bog and rocky outcrop. There, the watchful figures, in their subdued plumage of olive-gold sequins, standing each on its prominent green knoll or grey boulder, were a characteristic feature. Cock and hen would answer one another incessantly from their respective vantage points some yards apart with their mournful calls: *tu-lee*, *tu-lee*, *tu-lee* or an uneasy *squee-ee* from the cock, and a soft shrill *phee* from the hen.

Such a moor would also be well-stocked with curlew, lapwings and snipe, and was a favourite haunt of mountain hares; and on one I had the rare experience of approaching to within a few feet of a duck goosander — asleep on the green star-moss at the edge of a lochan, with the breeze ruffling up her dark-chestnut head feathers — before she sprang, croaking, into the water and slipped across it into flight.

The summer resident plover probably arrived in pairs, for a few pairs might appear in familiar moor and brae territories as early as the second week in February, and might even joy-flight spasmodically at that untimely season, though it was usually the latter half of March before there was any general joy-flighting. Then, the cock would circle at heights of up to 600 feet, while wafting around with a slow, tern-like flipping and high arching of wings, to the accompaniment of that melancholy, protracted *tirr-phee-ew*, *tirr-phee-ew*; but when joined by his mate his wing-rate would accelerate and he would break into a skirling *couerlee*, before plunging steeply down to the moor and reverting to the *tirr-phee-ew* whistling again. At times three would waft round together, *tirr-phee-ew*ing, with one keeping at a little distance from the other two. But a hard frost would silence all this response to the new season and environment, and the scattered pairs would flock together in the glens.

Within three weeks of their arrival in Badenoch the plover would be skirling desolately in the clouds above the snow-line on Carn Ban at an altitude of 2,500 feet, and one heard the lonely *phee-ee* and *phew-ee* of cock replying to hen, and knew that they had alighted and were keeping a watchful eye on the intruder. Before the middle of March they were on the high Cairngorm mosses, although at that date those were under total snow cover. Presumably these mountain plover only visited the high tops for so many hours a day until such time as the snow had melted from substantial areas of the mosses.

The parent plover became extremely secretive after the chicks had hatched, though the cocks would occasionally skirl in flight above their territories. By mid-July the summer's crop of young mountain plover were packing, and early in August one would find flocks of them in such high places as the 4,000-foot plateau of the Wells of Dee. Once, when I was watching stags in the Forest of Gaick in September, the pale underparts of a flock of young plover flashing in the evening sun caught my eye as they circled, twittering, over a 2,500-foot summit; and again in September, when searching for a snowy owl, I flushed a flock of nine from the summit of the Boar of Badenoch, five of which flew away westwards for as far as the eye could follow them.

All the resident plover had gone from the low ground in Badenoch by the end of August. Since we never saw any migrating through the strath by day, they probably did so mainly along the hill ridges and by night under normal conditions. Twice in one week in October, for instance, after orange and off-white highway lighting had been installed in Newtonmore and Kingussie, I was called to the window late at night by the clear whistles of golden plover. Rain was teeming down from a moor-level cloud base and, unable to navigate in the dense mist, they circled at roof height about the house, maintaining contact with their musical *thulee* whistles, which carried a special note of poignancy in such conditions. Some were still circling at first light the following morning.

Northern birds of passage apparently also migrated mainly over the hills, for in the middle of November one year I was astonished to flush a flock of more than forty from the snow-covered Sgoran, and these circled out over Glen Einich and alighted on the still higher Wells of Dee.

Although there might be more pairs of golden plover than dotterel in some areas of the high tops, the latter were the true mountain waders, for none nested on the moors below. During our first year at Drumguish there was a cold spell late in May, with a fall of several inches of snow. Nevertheless on the 23rd, when I was 3,000 feet up the stalker's path to Carn Ban, I perceived two plover-like birds silhouetted against the skyline of the col above, and recognised them immediately by their white head-stripes as dotterel. Their barred and laced wings and flanks put me in mind of red-legged partridges, as they fed daintily, with characteristic plover-runs along the edge of a snowfield, or stood fearlessly watching the dogs and myself only a few yards distant — *An t-Amadan Mointeach*: the Fool of the Peat Moss.

They were most inconspicuous on that hill-slope of dull fringe-moss broken up by dark-grey stones and boulders, which were blotched with black and silver lichens and purple cushion-moss. From time to time the brightly coloured hen arched tapering wings leisurely and gracefully, exposing their dull greyish-white undersurface. Her pale face and headband were a purer white than the cock's, and the crown of her head a darker brown. In the end they took wing together — the hen with a soft, slurred ticking note, not unlike the distant 'crackle' of a ptarmigan — and flighted in the erratic manner of ringed plover, revealing the rich chestnut and blackish-brown blazons on their bellies.

Elated at encountering a pair of dotterel in such an accessible potential breeding territory — Colonel Thomas Thornton had seen several on the Sgoran on 6 August 1796 — I climbed on up, happily unaware of the disappointments that awaited me in the months to come. On going over the summit of Carn Ban I observed a pair of golden plover accomplished by another small wading bird, a dunlin, for whose presence on Am Moine Mhor, the Great Moss, lying to the south and east of Carn Ban, I was not prepared, this being the first of its kind I had then seen anywhere in Badenoch. There were a few more dunlin on the Moss, and a pair were 'purring' and 'drumming' over the cold grey waters of Loch nan Cnapan, whose shores were still heavily wreathed with snow.

Six days later a dunlin was apparently feeding on numbers of blue-bodied craneflies, which were performing some peculiar rite at the spring on the summit of Carn Ban, dancing on their tails on the thick waterlogged cushions of black and golden moss. More dunlin were to be heard purring and *dee-dee-dee*ing, and plover skirling, in the mist-filled amphitheatre of the Moss; and a dotterel, reiterating a monosyllabic *peep*, interrupted from time to time by a strong trill, flew swiftly over my head, joy-flighting like a redshank — with wings momentarily still after a barely perceptible winnowing — before diving down to join its mate.

With May past I did not see very much more of dotterel, or indeed of any mountain birds, that first summer, for the weather in June and July was atrocious. To maintain seasonal continuity, let me pass on to my second summer, when the first three weeks of May were again wet, windy and very cold. However, I had not been on the Wells of Dee for many minutes on 21 May before I thought I heard the faint piping of a ring-ousel, repeated at regular intervals. Although ring-ousels had been recorded as high as 3,700 feet in the Grampians I could not credit that one could be singing at over 4,000 feet, and, on crossing the burn in the direction of the piping, I very shortly saw that it was indeed no ring-ousel that was piping. Between the snowfield on the steep west slope of the Wells and the sandy, shingly course of the spring with its grass-green edges, a flat of grey and pink gravel, particoloured in places with bleached yellowish-green tufts and grey and blackish-brown fringe-moss, sloped gently north to a scree of tumbled granite blocks and slabs at the base of Braeriach's summit peak. On the flat, a little way from the crystal waters of the spring, four dotterel with feathers hackled were running at one another pugnaciously, while reeling and trilling like turnstones fighting and displaying on the rocks amidst the sea spray. Although all four were brightly plumaged, the white markings were more extensive on the hens, and the blackish-brown colouring on their bellies was deeper and richer. They were, moreover, larger than the cocks and aggressively dominant to them. While I lay watching from a sandy bank at the edge of the burn, with my dogs sitting beside me, the four continually came together a few yards from me and separated again individually or in pairs. Now, the two cocks would charge head-on, and one, leaping over the other and pulling out a feather in passing, then immediately jumped on to the rump of a hen, which bowed forward until almost standing on her head, while he mated with her. Then all four would run swiftly apart, with frequent dips forward, ploverwise, to pick up grit or peck at a fox's dropping. Now, all four were still and uttering a soft and melodious, though staccato, *phwueep-phwueep-phwueep* (that ousel piping), or intermittently a thin linnetlike note, pitched so high as to be almost inaudible. Finally, the two hens, trilling loudly and with feathers hackled, ranged themselves side by side with a greater show of aggression than the cocks, until one broke off the engagement and ran at an attendant cock, chasing him away.

This was the first time that I had watched a species in which the hen took an equal, and even predominant, part in display and combativeness, and I was so fascinated that I was prepared to lie watching them indefinitely — it was tolerably warm for once — though I doubted whether I had ever witnessed birds courting in stranger surroundings, with thick cornices of snow still buttressing the drab-grey crag edges of the immense chasm below the Wells. However, after half an hour or so of intermittent display and fighting, all four dotterel gradually fed farther and farther apart and away from the burn, until I was obliged to rise and follow them, in order to keep them in view — when all took wing, with that reeling, whirring note, and vanished over the brow of the Wells.

This was the second year that I had found dotterel in two localities on the tops, and I was naturally eager to follow up their breeding cycle as fully as weather conditions permitted; but even as I had turned homewards, after the dotterel had flown away, a great cloud bank had loomed up out of the east and settled on the Wells, presaging the first long wet spell of the year — as if the tops were not already difficult enough to work on account of the strong winds that had blown almost continuously since April — and I was to see little of either dotterel or dunlin until 17 July. On that day two pairs of dunlin were persistently flighting with a pack of golden plover over the Moss, and when returning home in the early hours of the next morning I came across two solitary young dunlin about a mile apart. Shy little mites, they ran swiftly though hesitantly over the short grassy sward, one alone, the other with a parent flighting around.

A week later I was again on my way home from the Wells when my attention was arrested by a soft, spasmodic *pheep* that might have come from a golden plover. Orientating towards the sound, I very shortly espied first one, and then two dunlin-sized waderlings running erectly over the short sward. A mottled dark brown on their mantles, with some wisps of white down still adhering to their necks, it did not need the presence of the sober-hued parent cock to assure me that these little waders, with the bold black-brown bands reaching back from behind the eye to meet on the white nape, were indeed dotterel. Had I realised earlier that the nesting territories of the dotterel were hundreds of thousands of yards distant from their 'leks', I might have found other pairs in the corries west and north of the Wells. At first the cock took no notice of me; but, on the two young ones running off, he flew up with a purring *pyee-up* and pitched down again thirty yards away to perform the antic of rounding a nest in the grasses, half rising from time to time, to rock both wings in an injury-feint. Subsequently he flew back to his old place again, crouching down and nest-rounding, but not 'feigning', and, on my sitting down, tripped around me watchfully, bobbing occasionally and uttering a soft *thrrip*. As the evening was now drawing on I did not spend much time in attempting to find the young ones, which had disappeared while I was occupied with the cock, but continued on down the precipitous slope into Glen Guisachan, leaving the latter still tripping vigilantly along the skyline of that dry, south-facing pasture with its fine growth of grass and sedge, its club and reindeer mosses, bog-whortleberry and sparse growth of heather, now putting forth its creeping sprays of pink flower-buds — the highest in the western Cairngorms.

The traveller through the gloomy defile of Glen Guisachan — that almost subterranean pass through the Cairngorms from Glen Dee to Glen Feshie — hurried to escape from its oppressive iron-bound walls of cliff and scree and the fearfully broken, overhanging crags of the Devil's Point and Beinn Bhrotain; not that he could escape as quickly as he would have liked, for the pass was trackless and he must pick his way over the precipitous boulder-strewn slopes high above the impassable torrent-bed in the narrow and sinuous gorge below, trusting to deer paths to take him across the steep *allts* that rent one slope from the next. The roar of water plunging down a thousand feet from the crags was never stilled.

The last I saw of dotterel on the Cairngorms was on 9 August, when the Wells of Dee were more populous than on any previous occasion. Seven red deer hinds were grazing at the head of the springs, with three calves couched beside them. Solitary small tortoiseshell butterflies were constantly crossing the plateau in all directions. Of birds there were a golden plover with five juveniles, nine dark-grey young ptarmigan 'crackling' softly, a pack of ten pipits tossing themselves out over the chasm below Braeriach, and a juvenile dotterel making short flights from one place to another on the gravel, while uttering a repetitive musical *trrü*. Though it was boldly marked with dark stripes on the back, its white head-stripe was ragged, and the white band was barely perceptible on its pale, sandy-coloured breast. Two more dotterel subsequently went flying out over Loch Einich with their by now familiar erratic, curve-winged action; and another two were again accompanying the plover on Carn Ban.

Six years later I was able to watch the dotterel in their stronghold on the Drumochter hills, where there were certainly very many more pairs than anywhere else in the Highlands, and where there were also many more pairs of dunlin and golden plover than on the Cairngorms. It was mid-June, and heavy rain and cloud were boiling up the glens but missing us on the stony tops above 3,000 feet on one of the great hills on the west side of the Drumochter. For that mercy we were grateful, since we still had vivid memories of a previous visit to that inhospitable summit of naked slabs of schist and a sparse vegetation of azalea, crowberry, dwarf blaeberry, reindeer-moss and the soft fringe-moss. On that occasion we had been caught in our shirts in the most savage storm that I had ever experienced on the hills, and suffered a violent onslaught of almost solid sheets of cutting hailstones. I had never been colder or nearer exhaustion from exposure. Yet on our way up an hour and a half earlier we had sweated dry-shod across the lower burns; whereas on our way down we struggled thigh-deep through icy peat-brown torrents.

Lethal was not too strong a term for the sudden storms that could assail the wanderer on the high tops at any season of the year. However, on this 17 June the weather was kind to us and we enjoyed a most successful afternoon with dotterel on a ridge around 3,000 feet, finding no fewer than six clutches of heavily blotched, fawn or pale green eggs in neat little shallow cups, lined in some cases with stag's-horn moss, and either set among the bare stone or sunk in the fringe-moss. Five of the nests were approximately equidistant along a 1,000-yard stretch of the ridge. In addition a seventh nest contained two day-old chicks: exquisite yellow waderlings with bright brown makings, running hard all ways while their parent warned them with her soft musical purr as she lured us away. An eighth nest held one holed egg and two chicks just dry, and I had previously found a ninth nest with four eggs on this same ridge as early as 28 May. In every case except one only one parent was present, for not much was seen of the hens once they had completed their clutches. All the cocks were tame, and the one at the eighth nest, which had returned to the nest soon after we had found it, settled down so confidently that I was able to stroke its back. Another allowed me to lie stretched out and make a prolonged study of it at a distance of only three feet. Nevertheless, because most of the eggs were hard-set — two clutches were chipping — there was much injury-feigning with wings threshing, stiff legs rocking, and white-spotted tail, spread fanwise, brushing the stones. All but one of the nests with eggs held clutches of three. The exception had contained one egg in the morning, but when we passed it again four hours later I was astonished to find that the cock was sitting on two eggs.

On these bare western hills the dotterel were almost the only birds, save for the odd pipit and wheatear, a pair or two of golden plover and an occasional family of ptarmigan; but on the eastern hills, only two miles distant across the Pass, where the dotterel nested mainly in gritty 'pans' from which the winds had torn the sedge and fringe-moss, they were outnumbered by golden plover and dunlin. Indeed the numbers of the latter were sufficient to stimulate extravagant antics of luring and display, associated with an angry purring. There were larks singing on those hills too, which were vivid green with a thick vegetation of sedge and blaeberry, whose soft springy carpet the deer preferred to the rich green grass in the corries. Cloudberry and dwarf willow grew in extraordinary profusion around 2,000 feet, as did the delightful dwarf cornel with its four large white bracts and dark purple centre. Later in the summer, if there was sunshine, the lower slopes would be alive with thousands of mountain ringlets. Even the high tops carried a much fresher and denser vegetation than did those of the Cairngorms, despite the extensive areas of wind-eroded peat-hags and smaller pans of gravel and stones.

22 The Way to the Cairngorms

DURING our early years in Drumguish never a morning passed but my eyes first sought the Sgoran. Wherever else they might stray they always returned to the stalker's path that zigzagged up the huge bulk of Carn Ban Mor and disappeared over its brow. Never a morning but I speculated about the probable weather conditions on the high tops during the following twelve hours. And for all that time I never knew whether it was the mood of the explorer or of the naturalist that influenced me more, or alternatively the urgent need to revitalise life by periodic ascents to a place that was different. In the beginning the dominant attraction was the lure of the unknown. What kind of country was it up there? What lived and grew there? An additional attraction was that the high tops had been almost unaffected by any of man's activities at that time, for there had been no shooting on the Cairngorms, nor no more than a day or two's stalking in the autumn, for several years.

The fact that the hills were snowbound when we arrived at Drumguish only increased their attraction. I supposed, however, that the depth of the snow and the low temperature above 3,000 feet would render them impassable and impracticable for fieldwork; but I was keen to discover what fauna frequented Britain's highest mountain massif in winter.

Although it was the second consecutive open winter in Badenoch it was not until the end of March that the stalker's path had thawed out sufficiently to offer any possibility of my being able to reach the summit of Carn Ban, for the heaviest snow of the winter had fallen at the beginning of March. However, on a dull mild morning with a pall of rain-cloud swathing the upper 1,000 feet of the Sgoran, I left Drumguish at nine o'clock to try my luck.

The first stage was to the shepherd's cottage in Bailleguish; the second stage to the cable-bridge over the Feshie at Striontobair. Although at that point more than four miles of moor was behind me I had gained less than 300 feet in altitude. Before me lay a three-mile climb of more than 2,000 feet, which began on the very banks of the Feshie; then up through the small pine stand of Badan Mosach, with its chaffinches, coal tits and goldcrests, and a dipper bobbing on a rock in the smooth white slide of water that fell from top to bottom of the hangar of pines; and out on to the moors with their singing pipits, an occasional brace or two of grouse, and twenty-three stags.

Thereafter I lost all contact with the life of the glens, for the wide sweep of moor was silent and empty. No life, no sound except for the muffled roar of water tumbling down three *allts*, until I was in the clouds far up the path and confronted by an immense snowfield, which was three feet deep at its lower edge. I was then at about 3,000 feet, and the steepness of the ascent was approaching that of the roof of a house. For some time I cast about for a way through or round the snowfield, but with the cloud showing no sign of lifting it was evident that there was nothing to be gained by any attempt to climb higher. Reluctant, however, to turn back without a single record of mountain life, I stood listening, tensed to catch the least sound from the impenetrable grey cloud-mist swirling around me over the steep slopes, which fell away from my feet into the corries on either side of the path. And then, just as I was turning to make my way down the path's snow-filled trench I heard the mellow wooden 'rattling' of ptarmigan somewhere in the clouds above, and almost immediately afterwards the desolate skirling and *tirr-phew-er* of golden plover from below me. That was a bird I had not expected to find on the tops in March, and my frustration at failing to reach the summit of Carn Ban was lessened.

Ten days elapsed before the weather proved favourable for a second attempt to reach the high tops, and on a fine sunny morning I crossed the Feshie as the clouds were lifting off the Sgoran and packing away to the south. A lark was singing over the moors half a mile above the pine stand and twelve stags were couched on their favourite spur; but, with these exceptions, there were only grouse on the long climb to the summit, which I gained this time by leaving the buried stalker's path and striking directly up the snow-powdered heather on its north side. It was slippery going up the final almost sheer 700 feet, picking a route from one snow-free patch of granite grit to another; and when I had successfully negotiated what I supposed to be the ultimate ascent, I found that though I had apparently reached the top of Carn Ban, a seemingly limitless flattened dome of snow stretched away east and south, concealing what lay beyond. Immaculate snow, gleaming with ice-shine, mantled all visible land, and a dark prismatic halo lay upon the light misty cloud that partially veiled the sun.

I had been accustomed to the razorbacked mountains of Snowdonia, Cumbria and the West Highlands where, broadly speaking, one climbed up one side of a mountain and went down again on the other side. Thus I was not prepared for Carn Ban's immense summit plateau, and was at a loss to know what part of it to explore first. However, having been told that there were ptarmigan on Sgor Gaoithe, I bore away north-eastwards along the Sgoran, where the snow lay two feet deep. Half a mile farther on the plateau fell away steeply, before rising again to the wave-like crest of Sgor Gaoithe, which was a couple of hundred feet higher than Carn Ban. It was only after I had toiled up the broad snowfield to the crest that I perceived the cause of its curious configuration; for when I stood upon its little crag I saw that the eastern face of the Sgoran, all along its one and a half mile length, fell away from under me in one dramatic drop of 2,000 feet to the turgid black waters of a long glacial-formed loch – Loch Einich. On the far side enormous screes of a still higher mountain – Braeriach – swept down from its punchbowl corries even more precipitously into the loch.

This was a wilder scene than any I had then encountered in the Highlands. Colonel Thornton, who had adventured to the Sgoran twice in August 1796, and had even transported a boat to Loch Einich in October of that year, had also been impressed, as he had described in *A Sporting Tour*:

It is impossible to describe the astonishment of the whole party when they perceived themselves on the brink of that frightful precipice, which separated them from the lake below. They remained motionless for a considerable time equally struck with admiration and horror, the mountain above them to the right chequered with drifts of snow and differing but little from it in colour; the immense rocks to the left, separated by large fissures, the safe abode of eagles, and even the precipices around, appeared to them truly magnificent. Let the reader figure to himself a mountain at least *eighteen thousand feet* above him and a steep precipice of *thirteen thousand feet* below, encompassed with conical and angular rocks. Then let him imagine men and horses scrambling over huge masses of stones which, though of immense size, are frequently loose and at every step seem as if the next would carry them off into the air beyond its edges and the very idea would be enough to make him shudder. Yet the eye, having dwelt awhile on these frightful naked piles, is soon relieved, and feels an agreeable composure from the scene beneath, where the lake like a sheet of glass reflects on its extensive bosom all objects around. This lake, bordered by soft sandy banks, whose fine but partial verdure scattered over with small herds of cattle, grazing and bleating, and a single bothee, the temporary residence of a lonely herdsman, softens in some measure the unpleasant idea of danger which is apt to arise, while the solemn silence, interrupted only by the hoarse notes of the ptarmigans, increasing at the approach of the strangers, or by the dashing of the never-ceasing cascades, soothes the mind with the most agreeable emotion.

Cattle, herdsman and bothy alike had vanished, and as I had not set eyes at that point on any living creature I turned away to explore a further half-mile along the Sgoran to the Dubh Mhor; and a long plod it was through still deeper snow to that second crag of granite blocks. On the way, however, my dogs flushed two solitary ptarmigan and, to my astonishment, one or two meadow pipits. By the time I had climbed up to the crag the light cloud had passed from the sun, and it was hot enough to rest awhile and gaze down at black Loch Einich. Although there was considerable cloud at the horizon the prospect to all quarters was one of an endless massif, composed of snowy range upon range upon range, aligned one behind the other farther than the eye could reach.

Having traversed the length of the Sgoran, I was not inclined to trudge back through the snow the way I had come; so I went down its west face into the sheltered corries above Glen Feshie, where the snow had melted from large patches of fringe-moss and heather. Life was more abundant on those slopes, and during the first thousand feet of my descent the dogs flushed three brace of ptarmigan and three solitary ones, which went whirring and jinking out over the corrie, with the cocks rattling out their peculiar crackling notes and the hens crooning softly.

That was the last assault I was able to make on the high tops for three weeks, because of a prolonged period of westerly winds and rain.

23 Spring on the Cairngorms

THE SEVEREST blizzards of the winter might sweep the Cairngorms in March and April, and only the presence on the tops of pipits and golden plover confirmed the true state of the calendar. To complete my account of those months let me pass on to my second season in the Cairngorms. Although the winter had been exceptionally hard, the first snow had fallen early in November, with the result that there were no further storms after the opening days of February; and so violent was the subsequent thaw that I was able to make my initial ascent as early as 7 March.

It was not an opportune day, for persistent mist blotted out all visibility above 3,000 feet. However, there was always the possibility that it might be clear over the 5,000 acres of the Great Moss. As it happened, I never got down on to the Moss; but my day was not wasted, for when I was 2,500 feet up the stalker's path, which had been eroded and weathered by frost and spates into a *via dolorosa* of rubble and grit, I was astounded to hear the plaintive *phee-ee* of a golden plover. It was the more unexpected because at that early date not a single plover had returned to nesting grounds on the lower moors. The impression I had gained the previous year that the plover of the high tops were of a different 'race' to those of the moors was strengthened. My highest breeding record for a moorland plover was 1,500 feet. Between that record and those of the plover nesting on the tops existed a vacuum of some 1,500 feet in which none nested.

At the point where I heard the plover I also heard the musical trills of snow buntings and found them numerous thereafter in ones, twos and threes and in flocks of twenty-five. The cocks appeared almost pure white on the snow which, on the western slopes of the path was mainly restricted to broad tongues, many feet thick, extending like glaciers for a thousand feet and more from the tops into the corries.

Shortly before I reached the summit an eagle circled overhead, and then the clouds descended again; but after going over the top and a little way down on to the Moss I was soon brought up short by a snowfield, into which I plunged my long crook to the handle without striking bottom. Snow buntings, appearing as large as jackdaws in the white mist, hopped about the snowfield; three ptarmigan — snowy-white arctic birds with black-shafted primaries and seemingly the size of glaucous gulls — squatted at its edge. Not until I had approached to within a few feet of them did they begin to walk and then run over the snow, with their blackish-grey tails sticking out jauntily. A ptarmigan in the purity of its all-white winter plumage struck a new and bizarre guise at every encounter. I recalled my first meeting with one when, clambering over a granite ridge at a height of some 3,000 feet, I was shocked to see a white 'pigeon' with a prick of scarlet at the eye and a black fleck or two on the back standing beside a rock. At my questioning approach it strutted forward, opened its insignificant beak in a surprisingly wide gape, and twice insulted me with a deliberate, prolonged, rattling croak. And then, as it prepared to take wing, I noticed that its legs were clad with greyish feathers right down to the toes, and that when it keeled over in flight there was a black square in its white tail. One would have expected white ptarmigan flying over a white landscape to present a natural effect, but in reality I found the effect unnatural and incongruous.

Unable to perceive any bounds to the snowfield in the mist, I was obliged to content myself with these few observations; but turned back with the resolution that on the next attempt I would bring skis with me. This I did five days later on a cold morning with sunny intervals and the cloud ceiling at 4,000 feet, though by the time that I had shouldered my skis the three miles up Carn Ban I felt little inclination to ski anywhere, particularly as the snow, despite its great depth, was soft and slushy. However, once up, I made a speedy descent to the middle of the vast white waste of the Moss, which shelved down from Carn Ban and the Feshie hills and then, after a traverse of three miles, rose again to the 3,650-foot mass of Monadh Mor in the east and, in the north-east, to the whalebacks of Braeriach and Cairntoul, whose upper parts were hidden from me by a pall of snow-cloud. To the north the Moss terminated squarely and abruptly in an 800-foot drop to the chasm of Loch Einich.

Somewhere in the middle of the Moss there ought, I knew, to be a lochan — Loch nan Cnapan; but with every pool and peat-hag frozen solid and overlaid with snow, I failed to locate it and ended up in a maze of rocky cairns and pits that rendered the northern and eastern edges of the Moss almost impassable. On the way I saw a dozen or more ptarmigan and two solitary grouse, and on the 500-foot climb back up on to the Sgoran I heard other ptarmigan croaking and the trilled notes of a flock of fifty snow buntings near the wooden bothy above the spring on Carn Ban.

After these early expeditions phenomenal summer-like weather alternated with heavy rains, but there was much wind and some heavy falls of snow on the tops, and it was 8 April before I made another ascent. That too was a summer day in the glens, but a wickedly cold west wind was blowing on top. The snowclad summit of Braeriach was free from cloud at last, but west and north — from the Grampians to the Moray Firth — all was veiled in a fine-weather haze. With the snow melting rapidly below 4,000 feet and almost restricted to huge fields in the high corries and to narrow 'glaciers' down the *allts*, I was able to adventure as far as Loch nan Cnapan, some two miles down on the Moss. Heavy wreaths of snow still projected from the steep rocky banks of its grey sheet of water, three acres in extent. Nevertheless spring was well forward after the severe but early winter. Already the blue alpine foxtail grasses were barbed with yellow anthers; already the *Bombus lucorum* queens were zooming over the Moss on their mysterious errands; and pipits were in song everywhere from the ridge of the Sgoran down to the lochan. Of snow buntings and golden plover, however, there was no sign, and I was puzzled to account for the absence of the latter, unless it was that the heavy falls of snow had driven them temporarily from the tops. Throughout the four and a half hours that I was on the tops I saw only one pair of ptarmigan and a solitary cock alighting with characteristically tilted tail and shuffling wings. I lay down for a few minutes on the bank of a burn that meandered across the Moss from the Sgoran into the deep ravine of the river Eidart, and watched the ptarmigan, which were sixty feet away on the far side of the tongue of snow that filled the burn. Brown feathers, barred with black, were appearing in speckles and blotches on the hen's white back, while the cock — resplendent with scarlet eye-wattles — displayed a blackish grey silvering on the neck. Their tiny hooked beaks were perceptible only as points to their attenuate heads. The cock, strutting uneasily in and out of the rocks and hillocks and pecking now and again at the tips of the bents and sedges, kept up a constant, but barely audible, staccato *cr-r-ek* which, at moments of deeper uneasiness, would crescend into a full-throated 'crackle'. The hen, however, took very little notice of me, even when I struck matches to light cigarettes, and browsed on the sedge tips, only once uttering a soft clucking.

In the middle of April there was a cold spell with fresh falls of snow on the Sgoran, and a fortnight elapsed before a calmer and warmer day enabled me to undertake one of my longest and most gruelling days on the hills. I was out from eight in the morning until seven in the evening, and covered thirty miles (including thirteen on the tops) with only two fifteen-minute halts, encountering thirteen pairs of ptarmigan but only one pair of golden plover. On this occasion I went down past Loch nan Cnapan and right across the Moss, now a bleached yellowish desolation of sedges and bents extensively scarred with brown peat-hags. In its north-east corner I stumbled into a terribly broken country of outcrops and lochans and a chaotic jumble of boulders, where the predominant vegetation was moss and lichen: club and reindeer mosses, drab fringe-moss, cushion-moss blackening the grey boulders, and here and there small dark-green clumps and mats of crowberry. On the far side of this wilderness I came out on the boggy flats at the base of Braeriach and began the two-mile climb to the summit, 1,200 feet above. Traversing across the steep Horseman's Corrie, which separated the enormous southern boss of Braeriach from the most westerly whaleback of Cairntoul, I toiled up the latter's domed slope, ploughing ankle-deep through acres of shifting greyish-white gravel, which was almost devoid of vegetation above 3,500 feet, except for islands of desiccated bent and sedge, and here and there small clumps of moss-campion, whose cushions of minute thyme-like leaves were only just beginning to show green, though the few clumps 500 feet lower were wholly green.

Bearing west along the upper lip of Horseman's Corrie I found myself on the Wells of Dee plateau, which was composed mainly of gravel, much of its covered with an inch or so of fairly firm snow. The plateau sloped down to a central depression in which a frozen burn was just beginning to thaw out. The burn was fed by a number of springs — the Wells — bubbling up from the navel of the depression. Though snow and gravel were the dominant features of this arid, 4,000-foot plateau the high winds that swept its bare surface had prevented snow from drifting to any depth, except on the steep western slope of the watercourse. Here and there thawed patches revealed yellow-green whorls of map-lichen, contouring grey stones and boulders: spores and clumps of moss-campion as sapless and colourless as the sere fringe-moss surrounding them: and tufts of bent beginning to thrust out their short green blades. One patch of gravel bore the faint padmarks of a fox, but of animate life there were only two *B.lucorum* queens humming southwards and a solitary ptarmigan.

If I explored far enough over the western dome of the plateau I knew that I should arrive eventually at the edge of that mountain wall above Loch Einich, opposite Sgor Gaoithe; but what lay to the east? Slushing through the loose gravel I made my way to the lower eastern corner, where the burn disappeared, came suddenly to the edge, and looked down upon a stupendous scene. A colossal crater, nearly 2,000 feet deep, the Garbh Coire (Rough Corrie) shelved away to the gloomy canyon of the Lairig Ghru, which split the mountain masses of Braeriach and Ben MacDhui. The livid red screes of the latter swept down into the narrow rocky glen of the Dee, which was blocked off from the Rough Corrie by the overhanging and fantastically weathered black crags of Cairntoul and the Angel's Peak — more properly Sgor an Lochain Uaine, within the well of whose precipitous crags lay the Green Tarn at an altitude of over 3,000 feet. The greyish-green Rough Corrie, seamed with innumerable black burns and flood-channels, was a chaos of broken screes, granite blocks and craggy outcrops, of 500-foot stacks, sphinxes and dark-green 'cheese-rings', and heavily buttressed with snow, which receded in fluted tiers to the cliff-face, whose crumbling edge supported a fringe of snow cornices several yards in breadth.

Halting from time to time, while endeavouring to engrave upon my memory every detail of this fantastic corrie — soundless, lifeless, sunless; already in shadow by noon on Midsummer Day — I moved thoughtfully away and crossed the snow-lined banks of the burn at the point where it plunged over the broken edge of the plateau in a 500-foot fall into the corrie.

I was then some seven hours out from home, but I felt very fit and knew that there could only be one possible ending to such a day: to go down into Glen Einich and climb its east face below the crags of Sgor Gaoithe. The question was, could anyone other than a cragsman scale those apparently impregnable battlements? The map indicated a stalker's path which, however, terminated abruptly some three-quarters of the way up; so if in fact there proved to be no straightforward passage through the upper frieze of crags, I should have to retrace my steps all the way down to the north end of the loch and climb 1,600 feet up the Braeriach path — a descent and ascent of more than four miles — and thence round the north end of Loch nan Cnapan and through the wilderness. However, nothing venture, nothing win: first to traverse down the boulder-screes and loose gravel slides of Braeriach, with two out of every three boulders, overlapping one upon the other, rocking sickeningly under my feet; and then a straight descent through one vegetational zone after another: from the dwarf willow, bog-whortleberry and club-mosses, through the red, yellow and green mats of crowberry and sparse blaeberry and bearberry, and into the heather zone at about 2,500 feet.

In the latter zone I began to feel very tired, and remembered that I had neither eaten nor rested since setting out from Drumguish; so I sat down on a crag and examined the supposed path up to the Sgoran through binoculars, but could not detect any extension of it. However, wishing to find out what birds inhabited Glen Einich, and buoyed up by the thought of how short the way would be, once I had surmounted the Sgoran, I scrambled on down the watercourses and through the bushy heather, treacherous with hidden boulders, to land, thankfully, on the floor of the glen a little below an old bothy at the north end of the loch. All that was left standing of the bothy was a stone chimneystack, in the open fireplace of which some benighted hill-walker had recently kindled a fire. Surprised at seeing one or two ptarmigan at this low altitude, I made my way up the cheerless glen to the shingly beach of the loch and waded across a ford near the remains of old sluice-gates. There were a few grouse and pipits and four or five wheatears on the west side of the loch. There had also been a wheatear 2,500 feet up Carn Ban, and their presence was a promise of summer returning to the high tops.

Rain clouds were piling up over Braeriach as I trudged up the stalker's path, observing on the way a ring-ousel characteristically high up under the crags. About halfway up, at a point where the grassy slope became almost perpendicular, the path did indeed peter out in the boggy ground of a watercourse, though there were signs that it had at one time continued on up in the direction of a considerable snowfield below the craggy lip of the ridge; but as it was unlikely that I should be able to negotiate a soft cornice of snow clinging to an overhang, I cast my eye along the frieze of crags for an alternative outlet. At just one point immediately above me, and one only, a tiny crack in the frieze coincided with a straight climb, rather steeper than the roof of a house, up a green slope on which some forty hinds were sitting.

The chute looked practicable. If it was not, I could see no other way. So on and up, zigzagging from one side of the crag-bound slope to the other, to that cleft in the rock no more than twelve inches wide.

A last clutching scramble with hands and toes, and I was up and lying in a comfortable grassy hollow on the Sgoran, fanned by a fresh breeze, cool as spring-water after the windless glen. I was loath to rise from my couch, though there was not even an eagle to compensate me for my exertions: nothing but the great hills veiled by the pall of night cloud now settling upon them, and the gloomy abyss far below, its bottom hardly discernible in the fading light.

The painful descent of Carn Ban, knees quivering with fatigue, ankles turning on every loose stone, can be passed over in silence, as all in the day's work of a naturalist on the high tops.

24 Summer on the Cairngorms

SUMMER was long in coming to the high tops, and did not linger, extending only for a brief three months from late May to early August. The remaining nine months offered the naturalist barely enough fieldwork to justify such a disproportionate expenditure of time and energy in comparison with that afforded by the glens and pine forest. The first half of May had passed before the tops began to fill up, in species if not in numbers, with the summer breeding birds, and before the earliest mountain flowers were blossoming: in damp places masses of golden saxifrage and the yellow blooms of rooting marsh-marigold, on drier and grittier sites the just perceptible grey-green, lavender-like needles of dwarf cudweed and, side by side, the two lady's-mantles.

I was on the Sgoran for more than three hours on 10 May — which was a very cold and misty day after a prolonged period of windy and cloudy weather — and by the end of that time I had seen only one pair of golden plover at 3,200 feet and two eagles, one of which was perched on a crag of Sgoran Dubh, the site of a traditional eyrie. The remainder of my time had been passed watching half a dozen pairs of ptarmigan in their favourite corrie below the Sgoran and as many more on Carn Ban. Their 'machine-gun spitting' was to be heard from all quarters, and several were gliding buoyantly out from the slopes at a considerable height. All were in ones or twos, and paired cocks 'crackled' and postured aggressively at any intruder attempting to join them. Ptarmigan had long tails, and these they flirted almost vertically when they strutted, and also expanded them and half spread their wings when uneasy or when threatening a rival. The cock of one pair, after walking a few feet away from me, flew off with another cock, and on alighting on the far slope the two 'belched' at each other in different pitches. Subsequently one came gliding back on decurved wings a full half mile over the corrie. It resembled a black and white dove rather than a grouse, for by this date the ptarmigan were coming into full summer plumage, in which the cocks were dark grey with black mantles and the hens a mottled dark and light gold. In this plumage both harmonised perfectly with their background: that grey-green carpet of moss, broken here and there by bare patches of white gravel, which were strewn with grey boulders, stained purple-black, dark brown or bottle-green with pads of cushion-moss. True, I was watching my step on that treacherous carpet, composed mainly of a threadbare pile of soft fringe-moss, which concealed but did not shield the stones and boulders beneath; and keeping an eye on the dogs, in case they should flush any bird or beast, and on the weather and the distant prospect: but it was remarkable how often I put up the comparatively large ptarmigan almost from my feet. I watched one alight. Folding its white wings, it crouched close to the ground, assuming a curiously rounded posture, and was transformed into the likeness of a moss-clad boulder or a round lump of blackish moss, indistinguishable from other tumps and boulders; and as I approached more closely to it, so the bird crouched lower and lower. On the other hand, another cock appeared coal-black flecked with white, and was most conspicuous on the white gravel; though had I not seen it alight and squat down on that precise spot I might have mistaken it for a mossy stone. However, whether or not their colouring held any cryptic value, ptarmigan left a strong ground scent, which excited the dogs as keenly as that of grouse or white hares.

By 29 May there were several pairs of ptarmigan on the lowest part of the Great Moss at an altitude of exactly 3,000 feet. In the Cairngorms they therefore probably nested lower than the highest red grouse, since I had flushed the latter from a nest of seven eggs at a height of 3,200 feet.

Half an hour after noon on 17 June I was on my way up from the Wells of Dee to the summit of Braeriach. What little vegetation there had been on the plateau was soon all but swallowed up in a waste of grit, and that in turn disappeared in a zone of flat granite slabs, packed so closely, one overlapping another, that my way to the peak lay from one rocking slab to the next. On the topmost slab beside the cairn I sat down to rest in the sun, 4,258 feet above sea-level, and gazed out over the 1,500-foot precipices of the Rough Corrie to the dizzy crags of the Angel's Peak and Cairntoul. This was the wildest and most arid landscape I had then surveyed or even imagined. My horizon was restricted only by a haze at a distance of about seventy miles, which partially obscured the Moray Firth, Tarbat Ness, the Dornoch Firth and the mountains of Easter Ross to the north, and to

the west the peaks beyond Ben Nevis, the western sea, and perhaps
the Cuillins. But even on this sterile peak there was life: a fly and a
blue-banded hover-bee; a brown spider, with yellow stripes on its
abdomen, hurrying across a slab at my feet; and two solitary bumble-
bees, probably *B.lucorum*, humming past into space. Subsequently a
ptarmigan went 'crackling' off from the other side of the cairn, and a
few minutes later I heard a sharp *phwee-phwee-phwee*, and a dotterel
came flying over the Lairig Ghru and swept down to the Wells. After I
had stumbled painfully down again to the latter I found four dotterel
there: a pair and two solitary birds. The pair were quite fearless,
allowing me to stand and watch them at a distance of eighteen feet
while they tripped about with hackled feathers or flew up and alighted
again with a twittering purr or trill.

June was an ill-fated month. Continuous bad weather in both years
restricted me to only five expeditions to the tops, and four of those
were unproductive because of rain and cloud. Yet the period from
mid-June and mid-July ought to have been the most rewarding
season for a naturalist of the whole mountain year, for by the middle
of June there was a smooth, though still somewhat bleached, green
sward of bent and sedge over all the Great Moss and in the
punchbowl corries scooped out of the surrounding hills. Moreover
hundreds of deer now roamed over the tops, grazing on the new
herbage. Six or seven hundred could be counted at one time on the
Moss, which, together with the high corries of the encircling hills and
the heights above Glen Feshie, would remain their favourite pastures
throughout the summer, though small herds would range as high as
the Wells of Dee. For the most part they associated in sexually
discrete herds of from ten to a couple of hundred, though there was
usually a staggie or two with every herd of hinds and their ruddy,
dappled calves, and one or two big stags with large companies of
hinds. The remainder of the big stags, up to a hundred in number,
were always to be seen on the southern heights of the Moss, where
their antlered heads were clearly silhouetted on the skyline at a
distance of three miles. Although during those first two years I could
never afford time to stalk the herds, and approached the Moss openly
over the dome of Carn Ban, I usually descried them couched in the
peat-hags (much troubled by the flies even at that altitude) or grazing
in the deep green watercourses, long before they were aware of my
presence or of my dogs; and I often continued walking down to them
for several minutes before, at some distance between eight hundred
and one hundred yards, one or two of the hinds would display signs of

uneasiness, pricking up their long diamond-shaped ears in my direction. Only then would some drab-yellow old milk-hind with a ragged hide utter a lion-like cough. Her alarum would trigger off a pandemonium of hinds yelping and coughing, yearling hinds blaring like peacocks and calves bleating squeakily, as the various herds and families would converge spasmodically from all quarters of the Moss. Then the whole concourse of three hundred or five hundred, with the company of big stags among them, would canter across the Moss and make their way reluctantly down the steep slope into the gorge of the Eidart or slowly up into the corries high above the Moss.

There were occasions, however, when I would come suddenly over a cairn or down into a burn, and surprise a hind and a yearling. Unlike the other hinds, she would not trot or canter away after a preliminary stare, but would advance towards the dogs, stamping her foot; and if I sat down to watch her for a while, she would take up her stance on a nearby knoll, keeping me under observation. And there she would stand, uttering a warning yelp from time to time, while the yearling blared persistently, until I went on my way. She would have a calf lying up not very far away.

The return of the deer to the high tops coincided with the flowering of the mountain plants on the windswept mosses and plateaus. By mid-June there would be a pink flowerlet here and there on a green cushion of moss-campion, and the first pale pink stars of the creeping azalea, opening from their scarlet buds, replaced the minute, fleshy, purple-pink whorls of the crowberry, while the brownish-yellow, groundsel-like tufts of the dwarf cudweed (the Cairngorms' *edelweiss*) were just discernible, as they began their slow growth to their full stature of an inch or so. On the Wells of Dee, where the banks of the springs were becoming green, one could, if one looked closely, detect the diminutive blue-green leaves of dwarf willow and the dark green leaves, edged with red, of bog-whortleberry.

It was in the first days of July that the mountain flowering season attained to its brief zenith, and the boggier parts of the Moss were covered with spotted orchises, with here and there a butterwort and, still less commonly, though as high as 3,700 feet, the pale lilac-white marsh violet. On the banks of the watercourses with their thick green clumps of golden saxifrage, no longer in flower, appeared the alpine willowherb, the little viviparous polygonum (counterpart in form and colour of the lady's-tresses in the pinewoods), and in not more than a dozen places between 2,900 and 3,800 feet the exquisite star-saxifrage, which might be described as the London-pride of the mountains. Its fragile white star-flowers branched delicately from a slender reddish-green stem, six or nine inches long, which sprang from a leathery rosette of red, green and brown leaves. Each of the five white petals was stamped with two yellow or green spots. Those flowers with yellow-spotted petals were adorned with *small* orange anthers, those with green-spotted petals with *large* anthers.

In drier places there were many white flowers of cloudberry and here and there the small-flowered cow-wheat, of which some plants bore white trumpet-flowers, others yellow flowers (and purple-green leaves), and others again pale-yellow flowers with rose-purple tips. Additional examples of low-ground plants on the tops were provided by solitary specimens of chickweed-wintergreen, golden-rod, mouse-ear chickweed and mountain-everlasting, whose silvery leaves much resembled those of the alpine lady's-mantle, which was to be found as high as 3,600 feet; a dwarf form of eyebright and the heath-bedstraw also occurred at over 3,500 feet.

Above all it was the transient flowering season of the Cairngorms' one glory, the moss-campion which, by driving down an immense taproot, survived and indeed thrived on bare scalds of gravel among the less thickly piled boulders, and on flats and slopes of loose gravel, between 3,100 and 4,100 feet. It was only when I discovered that a cushion of campion nine inches in diameter might have a taproot more than twelve inches long that I appreciated the age to which a plant might attain. Wherever the alpine grasses, sedges and fringe-moss had failed to establish themselves, there could be found scattered spores and clumps and small islands of the campion. Where bare gravel extended down the slopes, there the campion might grow in comparative profusion as low as 3,200 feet, but on slopes with high grass little grew below 3,600 feet. On the Wells of Dee and the slopes of Cairntoul clumps of thyme-like flowerlets, varying from pale, almost whitish pink to the deepest china-pink of centaury, glowed against the sterile greyish-white gravel. So densely massed were the flowerlets, with their glistening white stamens, that they concealed the green cushions beneath. But so ephemeral were they that by the middle of July, when the early autumn of the mountains was already searing the tips of the bents with a bright chrome-brown, one had to search far and wide for a cushion still bearing flowers.

25 The Miracle of the Snow Bunting

BY THE MIDDLE of July in my second summer on the Cairngorms I had still not located any nesting dotterel or snow buntings. I had indeed not seen any of the latter since the first week in April, and had come to the conclusion that they had not nested in either year. However, I was determined to make a final search over all the tops west of the Lairig Ghru and Glen Dee. So, for the second year in succession, 17 July found me toiling up the stalker's path on a calm though cloudy day. There was a blue edge to the easterly moving belts of cloud that shed pale curtains of rain on the sunlit blue and green strath far below and, more ominously, fringed the Sgoran every now and again, swirling cold and grey into the corries and up the steep cleft of the Carn Ban watercourse. By the time that I had reached the summit and was making my way to the dank and dripping bothy for a smoke, visibility was down to twenty-five or thirty yards. Leaning up against the bothy door in the dead and heavy silence of blanketing cloud, the spectral thought passed through my mind — was I going to fail again? From time to time the oppressive grey nothingness enveloping me was relieved by the harsh cries of black-headed gulls hawking over the slopes of Carn Ban. The pall of cloud showed no sign of lifting; but I was in no mood to be defeated this time, and there was always the possibility that there might be clear interludes on the lower parts of the Moss. So, after finishing my cigarette, I shouldered my knapsack again and climbed up over Carn Ban's bare dome of fringe-moss. Only the gods knew what turned on that decision to go forward, for as I went down the grassy sweep of the hill on to the Moss the clouds began to lift off the surrounding heights, revealing at long last a marvellous clear green upon the hills and among the peat-hags. From all around came the *phee-ee* and answering *phew-ee* of watchful plover and the *whrrit-whrrit* of a joy-flighting dunlin.

With hope renewed I made my way over the Moss, where a stranger was a skylark uttering brief stanzas of song. My old friend the sandpiper was lisping on the shores of Loch nan Cnapan, where a black-headed gull floated idly on the waters, and I could see several gulls circling and hawking over the Wells of Dee.

For the fourth time I began the long pull up the steep eastern slope of Horseman's Corrie, bound for the Wells and, if there were no dotterel there, for Cairntoul and across Glen Guisachan to Monadh Mor — as yet unexplored. From the scree that covered most of the western slope of the corrie some small bird was uttering a faint, though persistent, monosyllabic *seep*; but I had gone out of my way too often before in this rough country, to be rewarded by nothing more significant than a fledgling pipit or wheatear, and took no notice of it.

It was then a little after noon, but I resisted the temptation to rest a while and have some lunch, though it was nearly always at that stage, when eleven miles out from home and with some four miles of climbing behind me, that I began to feel faint with hunger. However, I liked when practicable to reach the day's most distant objective before halting, and my goal was still 500 feet above me. This was, I reflected, my twentieth expedition to the tops in fifteen months, and I had still not found the young or eggs of dotterel, dunlin, golden plover or snow bunting . . .

My reverie was broken by a strong, sparrowlike chirp and a flash of black and white across the boulders before my incredulous eyes — God Almighty! Snow Bunting! I exclaimed in the stark amazement and hardly-realised triumph of that single second in time. But was this the snow bunting of the autumn fells and winter seashore? True, there was that same disclosure of broad white triangles on the wings; but this was a large jet-black and white bird with a white head or, alternatively, a pure white bird, except for the dull-coloured beak and round black eye, the jet-black mantle and tail. While making short flights from one big boulder to another, he repeatedly uttered a *tsee-eep*, like a reed bunting, and twice warbled a short sweet stanza of song.

But while I watched, fascinated, this unbelievable cock snow bunting in the full glory of his mountain plumage, that faint piping, which I had earlier ignored, continued nonstop from some point high up on the opposite slope; so when the cock crossed the burn to that side of the corrie — for he sang intermittently from various boulder stations up and down the corrie and several hundred yards apart — I followed him over and, while climbing about the scree, trying to locate the piping, became aware of his sober-plumaged mate preening herself on a slab of rock. She was familiar, with her drab-grey head, the brown spots on her cheeks and her ochre yoke and neck spot, her striped brown mantle and greyish-white underparts. From time to time she uttered the evocative winter trill, or the cock's *tsee-eep*, or a metallic whirring note, while the cock moved from one to another of his song-stations.

The latter was most beautiful when he perched with his back towards me, for then, from the base of his nape to the tip of his long tail, he was glossy black. The sudden revelation of white wings and white-edged tail, which now and again he spread and flirted like a redstart, was startling. No less startling to the ear, when he settled behind me, unnoticed, was the sudden brief phrase of lark-like song thrown into the silent corrie. Analysed, the song comprised a loud, clear, articulate and musical rendering of the reed bunting's slurred apology for one, and took the form of a low though distinct two-or three-note whistling prelude to a repetitive six-note phrase (of which the sixth note was pitched the highest), and terminated with a low chopped note. It was a jerky composition, which rose and fell in a manner pleasant to the ear, as he sat with legs flexed on his boulder, singing inconsequently, in the intervals of preening, without those laboured physical contortions characteristic of other buntings. Once or twice, when he flew up from his boulder and mounted to some height in crossing from one side of the corrie to the other, he uttered a longer and more sprightly and vigorous song-phrase that bore a still closer resemblance to the skylark's melody.

In the meantime the origin of that piping high up on the scree had eluded me for an hour or more, and neither of the buntings paid any heed to it, though the hen did not wander away from the scree as the cock did. Eventually I scrambled higher up the scree and, after a while, found the piper which, to my surprise, proved to be a fledgling bunting. Twenty-one days earlier I had gone up and down this same corrie, without hearing or seeing a bird of any kind in a gale-force wind and dense cloud. On that stormy day the hen would just have begun incubating. Had the weather been fair the cock would almost certainly have been in song. The fledgling bore no resemblance to the illustrations in any of the standard works. A plump 'corn bunting', its large-grey head had a tiny, conical, almost white bill and a black eye ringed with a bold white ellipse; there were dark brown stripes on its back and a whitish bar on the secondaries; its short square tail was edged with white; its belly a pale drab, the breast darker with faint rays and a dull whitish collar-line; its legs pale pink.

It continued to hop about the scree from slab to slab for some time, piping incessantly, without however attracting the attention of either of its parents, though once, when the cock dived headlong past it down the scree, it shivered its wings excitedly. But eventually the hen began collecting insects off the scree — probably craneflies, of which there were numbers of the copper-coloured kind about — and fed it from time to time in crannies among the boulders, into which it always retreated, on her approach, with a prolonged and emphatic whirring note, almost identical to that uttered by two young wheatears farther down the scree. Although the hen fed it several times in various crannies, the cock never did so; but when I left at 5.30 p.m., with cloud descending on the corrie, he was running about the mossy bed of a spring at the edge of the scree and picking up insects.

What more typical habitat for a nesting snow bunting could there have been than that steep jumble of tumbled boulders high up on the slope below the last remaining sizable snowfield in the western Cairngorms? The long strip of scree, some 3,000 square yards in extent, reached down into the pale green grass of the corrie. In spaces between the less densely piled granite-grey and pinkish slabs and boulders (weathered green with map-lichen and empurpled with cushion-moss) was the massed pink of moss-campion; hard-fern sprang from rock crevices, grey patches of cudweed covered banks of gravel, and the bed of the spring that trickled down through the scree presented a lurid patchwork of colourful mosses — green, yellow, brown and grotesque shades of blood: raw flesh, dull arterial crimson, congealed purple-black.

After watching the buntings for four hours I was content to call it a day. By this date some of the cock ptarmigan were packing in coveys of six or eight, leaving the hens to look after the cheepers. During the course of the day I had seen four broods of from two to seven cheepers. The clucking hens, shuffling drooping white primaries, would run around the dogs lamely, crouching close to the ground and feigning injury. In one instance two hens combined in decoying the dogs, though the older bird was much more persistent in her efforts than the younger. In another instance, however, when a brooding hen raised herself, clucking, her three cheepers quickly scattered, running swiftly with the aid of extended wing-arms. They were extremely obvious against the black and grey stones and gravel, for at this age their exquisite colour-pattern was gold and olive with a warp of brown threads. Even their legs and toes were covered with golden hair-like feathers, and the crown of the head was curiously marked with a dark brown lozenge with a pale centre. The hen, by contrast, was difficult to pick up in her blackish-grey plumage, as she walked calmly around or stood with her back towards me, unconcerned, only once shuffling her wings at the dogs. The cock, typically, flew over us once, swooped up with a belch, and disappeared.

The golden plover were also packing, and I was surprised to find on my way home in the evening one pack of eight and another of nineteen on the western slopes of the Moss. With them were two pairs of dunlin which, whether associating in pairs or temporarily solitary, persistently accompanied the plover in flight, even to the extent of one of a pair leaving its mate in order to take wing with a plover, before alighting again on a high knoll, from which it could obtain a good view of me.

When I reached home at 7.30 p.m. I had been out for eleven and a half hours. At 7 the next evening I made a sudden decision. I had awoken to a fine summer's morning and had passed a restless day sitting in the garden, getting up time and again to look at the Sgoran, while wondering if the weather was going to hold fair. I knew that there would be no peace of mind for me until I had visited the buntings again at the earliest opportunity, and that I would never forgive myself if I let slip the chance of hearing the cock's dawn song, though I feared that it was too late in the season for this to be at its best. Precisely twenty-four hours, then, after my return, I left for the tops again on a fine hot evening; but not without misgivings, for a breeze was blowing from the south, which was the rainy quarter in the Cairngorms.

I had never seen the green Dell of Bailleguish more colourful, with its glossy red, white and black cattle grazing peacefully in the evening sun and scarlet-billed oystercatchers probing the sward. Rabbits, running in circles on the braes, stretched their shadows on the green. The songs of larks filled the dell. At Badan Mosach the strong yellow rays of the sinking sun streamed through the pine trees, but thereafter the long climb up Carn Ban was a struggle all the way. However, by 9.30 p.m. I was leaning up against the ever-welcome wall of the bothy, looking out over a Highland settling into the peace of a summer night. Wind-streamered rose-pink bars and gold flecks of cloud still coloured the north-west sky over the Monadhliath, but westwards all was veiled: misty range upon range, cloud-layer on layer. Far below, the Spey was a burnished silver billhook. A raven floated slowly along the ridge, hooked head bowed. Already it was bitterly cold.

As I went down on to the Moss, with the light fading fast, I heard all around me the soft *phee-ee* and whispered *tissee* of unseen plover and dunlin, and herd after herd of deer were swallowed up in the concealing dusk, as they trotted away before me. There were more dunlin at the loch, and the reeling *durr-durr* of another came from the boggy flat at the mouth of Horseman's Corrie. Although it was then 11.30 the north still glowed a sullen red, but when in the end I sank down on a patch of gravel at the base of the buntings' scree and wrapped myself up in my raincoat, it was deep dusk in the cleft of the corrie below the steep lip of the Rough Corrie. Nevertheless, I could still distinguish the figures of feeding deer silhouetted against the pale northern sky.

The cold of the next four hours, while I hunched up on that bare knoll and subsequently against a wet rock, was unendurable, for the southerly breeze strengthened to a wind during the night, driving the clouds into the corries. Their dank rain-mists swept into and up Horseman's Corrie, as I lay looking up at the dim stars of the *Plough* and the *Pole*, and when a perceptible lightness was stealing into the corrie I could bear the cold no longer and rose reluctantly to make my way homewards. But as I stumbled stiffly down the corrie I noticed that a cloudlet over the snowfield was flushed with the brown of dawn; so I turned back to wait a little longer, and saw deer begin to appear over the eastern slope. Ten minutes later a ptarmigan 'crackled' from that quarter, and more loudly five minutes after, and the new day had begun at long, long last. There followed another chilly interlude of stamping about from one foot to the other before my benumbed senses awoke to the realisation that the cock bunting had uttered a single faint phrase of song from a point high up beside the snowfield. That was at three o'clock, and in another ten minutes he began to sing in earnest at half-minute intervals from various boulder song-stations, but did not, to my disappointment, joy-flight. Although I waited for another half-hour or so he did not alter his routine, and since there was no sign of his mate or the fledgling, I abandoned my vigil and left him still singing.

26 The Snow Buntings again

FOR A FORTNIGHT after that night out in Horseman's Corrie I was strathbound with bad weather, and I had to wait until 24 July before a fair, clear, cold day dawned.

When I reached the Corrie at noon all was quiet, and it appeared to be occupied only by two broods of fledgling wheatear twins. However, I lay down for a snack of lunch at the bottom of the buntings' scree, from which the snowfield had now vanished, as had the big field on the summit of Carn Ban a week earlier. Indeed the only snow now remaining from the winter storms were a few small patches in the semi-permanent fields on the western precipices of the Rough Corrie.

After I had been lunching for ten minutes the hen bunting suddenly darted past me from the scree and pitched out of sight on a flat higher up the corrie. When she did this twice more at five-minute intervals I guessed that she was either building a new nest or feeding another brood of young — unlikely as that seemed at such a late date — and I soon located her hopping about a little scree of small boulders and large stones. Since she ignored my presence I sat down twenty feet above her, and after peering down a crevice in the boulders for a minute or so she popped in. I allowed her five minutes and then, as she did not reappear, stepped over the slabs to investigate; but though she then emerged, I failed to find any nest. So, marking the place, I retreated and let her dive in once more. On flushing her again, I located the nest a foot down under the boulders. To my surprise and delight it was filled with a soft shapeless heap of dark grey-brown down, comprising, so far as could be determined in the dimly-lit cavern, from four to six nestlings. From time to time a faint *peep* came from the amorphous mass. The rounded nest, constructed of dead grasses with a few small white down-feathers woven into them, was raised up out of the wet black trodden soil under the slabs and had evidently been built on a foundation of one or two earlier nests. It was

not difficult to understand why this was only the ninth nest of snow
buntings to be found in the Cairngorms in the fifty years and more
since the first eggs had been taken.

Subsequently I watched the hen entering the cavern with small
beakfuls of craneflies, with which she fed the nestlings for five minutes
or more at a time. There was no sign of the fledgling of the first brood,
though the cock passed silently over the scree once and later whisked
over the brim of the corrie above it. Although the hen continued to
visit the nest without any trace of uneasiness, I was not prepared to
take the slightest liberty that might jeopardise her chances of
successfully raising this second brood; nor did it seem likely that I
would observe anything of interest at this early stage. I decided,
therefore, to make my way over to Cairntoul, with a view to
continuing my search for dotterel, and with the possibility in mind
that I might also find a second pair of buntings. But there were only a
couple of families of ptarmigan on those wastes of gravel, boulders
and screes of granite slabs bare of vegetation. The cheepers whirred
away with the cocks into the Rough Corrie, while the hens stayed
behind, shuffling their wings or standing watchfully on boulders.

All the way home across the Moss I was accompanied by a pair of
answering golden plover, as the cock continually moved ahead and to
one side of me, while his mate followed behind me on the other side. A
second pair were accompanied by a full-grown golden-brown
fledgling; but they were the only plover I had seen all day, and I had
recorded none on the lower moors for the past two weeks. A second
mountain nesting season was drawing to a close; the dunlin had gone,
no black-headed gulls had visited the tops for a week or more, and
though some pipits were still in pairs others were flocking, and of
eight wheatears noted during the day only one had been an
adult —a cock.

I visited the young buntings again on 30 July when, assuming them
to have hatched from twenty-four to seventy-two hours before I
found them with eyes tightly sealed, they were between seven and nine
days old. For once, I was fortunate with the weather, for although
cloud lay on the tops from 3,000 feet upwards until an hour after
noon, it lifted with a cool north-west wind ten minutes after I arrived
at the buntings' scree, and I was able to lie in the sun and watch in
comparative comfort for two and a half hours. A small tortoiseshell
butterfly flew past me once, and four or more fledgling wheatears and
a pair of pipits hovered inquisitively above the scree. At this stage
both cock and hen bunting were bringing food —exclusively

craneflies – to the nest with impressive regularity. The hen again took little notice of either the dogs or myself, thirty or forty feet from the nest-site; but the cock (who was beginning to moult) often came and went a number of times, fluttering over the scree like a beautiful black and white butterfly or sweeping in circles around it, before alighting on a boulder, two or three 'flutter-hops' from the nest-cavity, and announcing his arrival with an emphatic *phwuee* or soft trill. Even when standing on the slab beside the cavity, he might first retreat two or three times before finally diving in with his beakful of craneflies. However, he fed the nestlings on eight occasions – in comparison with the hen's nineteen – during my period of observation, while the hen also made three visits to the nest without delivering food: once when she found the cock already in the cavity, and twice when she dropped down to bring out a faecal sac. Both, on average, took between two and four seconds to deliver their loads; but in one instance, when they arrived at the nest-slab together – though from different directions – the hen went in first and stayed down for eleven seconds before emerging and soliciting the cock, who surrendered some of his craneflies. She was in and out of the cavity with these in a couple of seconds, and again postured with arched wings to the cock. This time, however, he refused her and, when she flew off, entered himself, but came out without feeding the nestlings. Yet one minute later he accompanied the hen in and emerged after three seconds, followed by the hen carrying a sac. After each visit both flew away immediately in opposite directions to new collecting places within a seventy-five yard radius of the nesting scree, picking up craneflies from boulders or crannies, but never on the wing. Sometimes they flew across the scree for a final pick-up on the opposite slope before coming in to feed the nestlings.

At every visit by a parent the nestlings could be heard uttering that prolonged whirring note, which the earlier fledgling had also emitted when being fed. The latter, now strong on the wing, visited the scree two or three times. Although three weeks had elapsed since I had last seen it, there was no noticeable alteration in its plumage. Towards the end of my stay I went over to look at the nestlings. They still formed a conglomerate heap of dark grey down, and it was only possible to distinguish the yellow external gapes, horn-coloured beaks and small black eyes of three separate entities. From such evidence as was available in ornithological literature I estimated that they would not fledge before 3 or 4 August at the earliest, and I determined to revisit them on the 3rd.

Again I was fortunate in a spell of really hot summer weather; but I was not to enjoy the full benefit of this, for a mist descended on Horseman's Corrie immediately I entered it. On arriving at the nesting scree at 11.45 I found the pipits and two or more young wheatears again present, perching inquisitively close beside me and even on the slab above the nest; and while I lay watching the buntings, twelve deer — seven hinds, four calves and a staggie — grazed down the west slope and finally couched a few hundred yards below me. They remained there, chewing the cud, throughout my stay, now hidden in the mist, now revealed when it lifted momentarily.

Both bunting parents were working feverishly, and during the first half of my vigil the cock kept pace with the hen, each carrying ten beakfuls of food to the nest. The cock indeed arrived twice in thirty seconds during one burst of six visits. On two or three occasions both arrived at the scree together — though from different directions as usual — and on one of these the hen solicited the cock, who did not respond to her advances. Thereafter the cock disappeared, while the hen made ten further visits.

While watching their activities I had found myself wondering why, when one pair of snow buntings could feed and rear a proportion of two broods of young in less than two months on a superabundant supply of craneflies, a nesting tradition had not been permanently and more numerously established on the Cairngorms among their race. What peculiar problems did snow buntings have to overcome in nesting and raising young? It was difficult to assess the effect of the mountain climate, though the short summers and unseasonable snowstorms did not prevent wheatears and pipits from nesting annually in the same habitat. Of possible predators I could think only of foxes, though possibly a stoat might visit the highest tops once in a while, and even as I pondered the question a small peregrine tiercel suddenly shot out of the mists and perched on a boulder a little above me. At its appearance the cock bunting ceased hopping about the scree in search of craneflies, perched on a boulder uneasily, and said *tew-ee*; but when I stretched out my hand to reach for my binoculars the tiercel sheered off into the mists again. However, the scarcity of nesting buntings on the Cairngorms was probably not due to any climatic or predator factor, but to the geographical factor that Scotland was an impermanent southerly extension of their boreal breeding range.

After an hour or so I went over to inspect the nestlings and found only two in the nest, with a third perched on its rear rim. When I squinted down at them through a chink in the boulders the latter disappeared into a little cavern behind the nest. However, while I was searching for the place in which the hen dumped the faecal sacs, she entered the nest-cavity again, and when I looked in a second time there were *four* nestlings in the nest! I estimated that they would not emerge from the nest-cavity for another three days, and that in any case I could not fail to see them, above or below ground, if I visited the corrie again on 6 August. In the meantime there was little to be gained by prolonging my stay, since the mist showed no sign of lifting; but as I breasted the western slopes of the Moss the treacherous mists did lift, and I turned to look back — as I had so often done before — at the long green Horseman's Corrie, flanked on the west by the tremendous boss of the Wells of Dee, and on the east by the grey *couchant* sphinxes of Cairntoul. Two days rest, and I would be plodding up the steep burnside once again to that small scree almost at the head of the Corrie, for the last act of the drama. The mountain gods smiled.

On the night of the 5th a violent rainstorm deluged the Grampians, and for the next three days it was intensely cold, while rain-mists blotted out the moors on the very threshold of Drumguish. Instead of the 6th it was the 9th before I was able to climb to Horseman's Corrie, where I found that there had been a fall of fresh snow on the site of the old snowfield at 3,800 feet. This new snow melted away during the afternoon, though traces of a frieze remained here and there on the western lip of the Rough Corrie, which still retained two sizable fields of 'permanent' snow, as did the east face of Braeriach below the summit cairn, and the west face of Ben MacDhui. Nevertheless, it was the warmest and sunniest day of the summer, and herds of deer, three hundred strong — some including both hinds and stags — were grazing in all the high corries, though the company of a hundred big stags was as usual on the favoured southern heights of the Moss. At Loch nan Cnapan I flushed a brace of grouse and the peregrine tiercel again, and at 3,000 feet on the bog at the base of Horseman's Corrie I overtook a three-quarters-grown frog. Wheatears were present as usual in the lower reaches of the Corrie; but as I began the steep ascent up the burnside I had a presentiment that I was not going to find the buntings at the scree: the torrential rainstorms would have drowned the nestlings, and the parents would have left.

The scree was indeed silent and deserted. Even the pair of pipits had gone. But when, filled with foreboding, I lifted away the boulders from the nest-cavity (for the first time), I found, not drowned nestlings, but only an addled egg half buried in the trampled nest platform. It did not seem credible that, within five days at most of emerging from the cavity, the nestlings could have flown; so for the next three and a half hours I searched far and wide over the corrie and the Wells of Dee. But there was no sign of snow buntings, young or old. Once again the mountain weather had frustrated me.

To conclude this account of my quest for snow buntings I must go forward two years, when although deep snow was still lying in Horseman's Corrie at the end of May one or more buntings returned before 8 June and I was able to make eight visits to the corrie between that date and 30 July.

It was so cold in the cloud-swathed corrie on 8 June that I found it impossible to linger there for longer than forty minutes, during which time I glimpsed a cock whirling about the snowfields on the east side of the corrie; and although I spent more than two hours in the corrie on the 19th only the cock was present and singing fitfully from the snowfield on my arrival, but thereafter silent; while during a stay of more than four hours on the fine afternoon of the 25th I saw no buntings at all. On 7 June, however, which was another very cold and windy day, two cocks were singing from screes on opposite sides of the corrie. On a hen appearing, there was a show of aggression between the cocks, one of which spread his tail to her and picked up a beakful of the seaweed-like *stricta*, before all three disappeared. Later I found a cock singing on the summit of the Angel's Peak above the corrie, and here there was a fledgling, which must have hatched not later than 15 June. Although one cock subsequently indulged in brief, low shuttlecock song-flights, none of the buntings collected any insects, and since the hen was not seen again, she was presumably incubating a second clutch of eggs.

July 14 was a fine summer's day, but no buntings were to be seen or heard when I reached Horseman's Corrie at 11.45, and it was not until I had climbed over the crest of the corrie towards the Angel's Peak that at 1.15 I found a cock singing from various stations over a wide area of this new locality, though no hen was to be seen. On my return to Horseman's Corrie an hour later the other cock was persistently chasing young wheatears, when not singing from his accustomed scree or less often while making short flights on half-spread wings; but here too there was no hen.

On the 19th, which was again warm though windy, the Horseman's cock was in fitful song at 11.30, and after an hour and a half I located the hen and the now silent cock feeding among the bents; but I was unable to trace the hen back to her nest. Five days later the feeding cock was accompanied by the fledgling, but the weather was too bad to keep track of either. Finally, on the 30th, I found the fledgling, then not less than twenty-three days old, feeding by itself; but though it was a calm warm day no adults appeared during a stay of two and a half hours.

Three years later, on 3 August, I watched the Cairngorm snow buntings for the last time. It was as if I had returned home, for once again a pair of ptarmigan, accompanied by a single cheeper, were near the bothy on Carn Ban; once again there was a dipper at the head-waters of the Eidart, a little below 3,000 feet in the middle of the Moss; and once again there was a pair of snow buntings in Horseman's Corrie, though a kestrel on the Angel's Peak was a stranger.

The two buntings swung across the corrie in opposite directions as we arrived, and that was the last we saw of the hen, who was probably busy with a second brood. Later, however, we watched the cock feeding a single fledgling at regular ten-minute intervals. The fledgling, which must have hatched not later than 21 July, had probably not been on the wing for more than twenty-four hours, and remained somnolent on or among the boulders during the cock's absences, but responded with a soft cheep and shivering wings to the latter's approaches with food. The cock was collecting all his insects on the ground, and though they were winged (and appeared black) may not have been the customary craneflies, which were not as noticeably abundant as in earlier years on the high tops. In the intervals of collecting food he engaged in the familiar short flights or perched on boulders, but did not sing; and in the end he called the fledgling away from us, rousing it to run very nimbly downhill with occasional short flights.

The main attraction of the Cairngorms to snow buntings was undoubtedly the presence beside suitable nesting screes of those semi-permanent snowfields — which always held snow until mid-July or early August — and their derivative melt-water and springs, whose mossy beds drew large numbers of craneflies. To such sites the buntings returned year after year, and although it had been stated that there were summers when no buntings nested on the Cairngorms, I would wager £1,000 that, given reasonable weather conditions, I would find them every July in Horseman's Corrie or in one of the adjacent corries *if* I could get myself up there.

27 Autumn and Winter on the Cairngorms

BY THE MIDDLE of September it was full autumn on the tops, and the old enemy, Braeriach, was a brindled tawny, grey and yellow-green. Pipits still lingered up there and, to my surprise, a pack of twelve golden plover, of which the majority were juveniles, were on their favourite heights near the Carn Ban bothy. Although a few scores of hinds and followers were scattered about the Moss, the herd of big stags had dispersed to their rutting grounds; and when early in October I set foot on the giants of the eastern Cairngorms, stags were wailing and snarling in the corries and on the high plateaus on both sides of the Lairig Ghru, while lone wanderers, roaring disconsolately, were crossing the pass from east to west, by way of the spacious grassy plain between Ben MacDhui and Cairn Lochan.

It had been a superb Indian summer's morning when my companion and I left Loch Morlich and made our way through the Rothiemurchus Forest, with its chaffinches, coal tits and black grouse, and its black sympetrum dragonflies, to that traditional pass through the Cairngorms, the Lairig Ghru. Although the latter was the old drovers' route from Speyside to Deeside I found it hard to credit that droves of a thousand head of cattle could have ever have been driven through so steep and narrow and rough a defile. Long stretches of it were composed of broken rock and scree, though no doubt an immense quantity of debris had fallen from the Lurcher's Crag (Creag an Lethchoin) and Braeriach during the hundred years or more since the men of Rothiemurchus had gone up the pass every spring to clear a way through the chaos of scree and boulders. Where the pass climbed above 2,000 feet, on approaching the watershed, cattle could only have negotiated it in single file and, even so, not without a broken leg or two. Possibly the larger droves followed the longer route through the adjacent Lairig an Laoigh and then onwards into Glen Derry.

We, however, had left the track, once we were clear of the forest, in order to climb the 3,450-foot Lurcher's Crag. This was a jagged ridge strewn with the ubiquitous granite boulders and gravel, and thickly carpeted with fringe-moss, extensive patches of bright green crowberry and, here and there, trailing red, purple and brown clumps of azalea, still bearing their pink flowers. Reindeer and club mosses and the flame-tinted bog-whortleberry completed the vegetation on the ridge, though from 3,750 feet upwards on the slopes of Cairn Lochan the dwarf willow was conspicuous in its autumnal gold.

Of birds there were three snow buntings — probably a pair and their fledgling, for one was a cock in autumn plumage — a solitary eagle, and a pack of more than fifty ptarmigan rising into the sun like white doves from a jumble of grey boulders.

From the summit of Ben MacDhui (4,296 feet) — a plateau of gravel, boulders and granite slabs — the visibility was for once almost perfect, and the full horizon of more than ninety miles may have been within our ken; but although it had been really hot on the Lurcher's Crag an icy gale was blowing on MacDhui, and we were glad to shelter in one of the roofless surveyors' bothies, before making our way down the northern slopes to the watershed above Loch Avon, with its molten sands, black rocks, precipitous waterslide, 'sawn-off' head and huge stacks of 'cheese-rings' rearing up on either side.

And so home in the shadow of Cairngorm, through the steep Coire Cas, where a big stag was rounding up his thirty-five hinds and followers, to a still summerlike night in Glen More and the cold silvered sheet of Loch Morlich, stained saffron by the last glow of the sunset fringing the Monadhliath.

Towards the end of October a broken frieze of snow hung from the western lip of the Wells of Dee; but it was still autumnal on the high tops, and the Great Moss presented that tapestried panorama that made this the richest colour-season of the year on the moors and in the glens: a tapestry of black peat-hags and pale tawny deer-pastures, with a black-green shading of mosses, lichens and crowberry on the grey outcrops, and here and there patches of dark lake-red or rich orange-brown sphagnum and scarlet spots of *cladonia* lichens.

Although the Rut was drawing towards its close, the lowing and belling of stags — like the wind sobbing through the glens — carried faintly across Loch Einich from the corries of Braeriach; but not a roar was to be heard on the Moss, despite the fact that there were still some scores of deer up there, including one herd of twenty-nine hinds and followers with six big stags, another of twenty-eight with one small stag, and a few wandering stags.

The last of the summer resident birds, the pipits, had gone, though one or two grouse were evidently going to winter with the ptarmigan on the Moss until the snow buried all vegetation. There were numbers of ptarmigan on the rocky northern edge of the Moss and in the eastern corries of the Sgoran, from which they sailed out high over the turgid black-brown waters of Loch Einich in sweeping arcs with not a flutter of their fingered primaries until far out in space. I must have seen forty or more in a couple of hours, the majority in droves, but a few in pairs or solitary. One or two snow buntings had come in to winter, and there were more eagles to be seen than at any time previously; for when I was approaching Sgor Gaoithe, first one eagle and then a second — a young bird with a broad black bar across the base of its tail — glided along the weatherside of the Sgoran, occasionally hovering without perceptible wing movement. Later, when I was sitting in the lee of the *sgor*, a third coasted along the edge of the crags, and three more were sweeping and hovering over the Moss. All were, no doubt, hunting for ptarmigan.

It was still the season of autumn migration. On the way up Carn Ban I noticed that the path was stained everywhere with the purple-black residues of berries, and at about 2,500 feet a flock of several hundred fieldfares, accompanied by some redwings, rose from the slopes above. As I approached Loch nan Cnapan I heard geese music, and over the western shoulder of Braeriach, at a height of 3,750 feet, came fifty-three grey lags in four gaggles, which merged into three when they passed low over the lochan. Their loud cheery gobbling and honking caused quite a commotion on the Moss: packs of ptarmigan 'crackled' and the deer looked up at these noisy travellers, turning their heads right back over their shoulders to follow them on their way.

Although stags were still roaring from the Braeriach corries above Loch Einich in the middle of November there were no deer on the Moss, which was thinly covered with a sparkling mantle of snow. With a hot sun in a blue sky from sunrise to sunset, and a still, chill air, the day was as perfect as one could hope to experience on the tops. A flock of forty snow buntings were feeding on a snow-free slope above Loch Einich, and the light covering of snow on the Sgoran revealed unexpected life. There were several tracks of hares, which the dogs had seldom put up on the tops. For that matter, I had not seen more than fifty hares in two years in Badenoch, and never more than four in one day. Yet ten years earlier they had probably been more numerous in Badenoch than in Argyllshire, where I had counted a hundred in a single morning. More remarkably, there were also tracks of stoats. One track led up to the cairn on the Sgoran at a height of 3,635 feet. Possibly the stoats had followed the hares' trails, for I had not previously encountered them higher than 1,200 feet, where they were generally to be found near rabbit warrens, 'snaking' along the runs, examining one burrow after another, or popping in and out of stone dykes. On one occasion I watched a pure white ermine poking its head in and out of the stone wall of a sheep fank, characteristically curious about my two dogs. However, when one of them went right up to the hole, the little fury chittered at him with an explosive squeak, before withdrawing into the hole, from which it was not to be lured again. The few stoats that one did see in Badenoch were in a variety of pelages, from the snowy-white winter ermine — faintly yellowish on the chest at a distance of a few feet — to those parti-coloured individuals with brown faces and white ears (edged with black), white throats and bellies, and brown backs with a spot of white on the rump. In March a pure white ermine, weaving sinuously over a bleached tawny bog, and stopping every few seconds to stand erect for a 'look-see', was as ludicrously conspicuous as a mountain hare on the same terrain.

It was an emotive experience to leave the Sgoran in the heat and light of day — for it had been hot enough to enjoy a snack of lunch while sitting on a boulder – and, later, look up from Glen Feshie, 2,000 feet below, and see a rosy glow on the white tops; then cycle across the moors, with the grouse calling, into the smoky glare of an orange sunset, and finally go down the hill to Drumguish in the near-darkness, with the lights twinkling in the strath – so short was the Highland day from November to January . . .

Short indeed, when in the middle of December the sun was only just rising over the Feshie hills as I climbed up above Badan Mosach. I was at the bothy by 11.30, when I was immediately enveloped in a snowstorm, which covered the upper 400 feet of the Sgoran. However, I groped my way down on to the Moss and, on approaching Loch nan Cnapan (now frozen solid), emerged from the clouds. There had been considerable changes on the tops since my previous ascent five weeks earlier. The upper parts of the stalker's path were filled with long troughs of frozen snow, packed several inches deep, which would probably not melt before the spring thaw, and a wall of snow reached almost to the roof of the bothy. On Sgoran and Moss the crust of frozen snow, an inch or two thick, broke at every step, though the snow cover thinned out in the lower parts of the Moss, revealing sprays of crowberry, the thrusting fingers of fir club-moss, and brilliant red tufts of sedge and bent; but the bird life comprised only a few individuals of the three winter-resident species — a score or two of snow buntings, two solitary *go-back*ing grouse, and some ptarmigan 'reeling' persistently. The latter included the Carn Ban pair, which had already taken possession of the snowfield below the summit.

By 10 January there had been little further change, though the upper part of the stalker's path and the adjacent slopes had disappeared under a spread of iced snow, on which it was difficult to keep one's feet, with the alternative of sliding steeply down into the corrie. In the glens it had been the mildest of days, but a strong wind was blowing on top and it was so cold at the bothy that the dogs' coats became frizzed white with hoar-frost. There were a dozen snow buntings near the bothy. What possible attraction could the Cairngorms' inhospitable tundras hold for these small passerines, barely able to keep their balance against the blast while pecking at tufts of sedge protruding from the snow?

With mist reducing visibility to fifty or a hundred yards there was nothing to be gained by venturing down on to the Moss or along the Sgoran, so I slid down into one of the Glen Feshie corries, in which there were both grouse and ptarmigan, and went home through the pine forest, with its redpolls, bullfinches and chaffinches, and sixteen big stags couched for the night under the trees.

Six days later I was luckier with the weather and succeeded in reaching Sgor Gaoithe on a cloudless day with a scorching hot sun, despite a very hard frost. Although a hind and her yearling were 3,000 feet up on one of the Glen Feshie hills, the Sgoran was for the first time in two years devoid of all life. There were not even tracks of hares or stoats. Below Carn Ban, however, there were more than a hundred snow buntings, and five ptarmigan were walking, running and surprisingly sliding and falling on the frozen slopes, in the intervals of pecking at the few perceptible bents. A noisy pack of thirty-five grey crows were quartering the moors below the snowfield, cawing their trenchant *quor-rah, quor-rah, quor-rah.*

On the morning of the 26th a black clump of more than a hundred stags could be discerned on the white face of Carn Ban, and later in the day I found that I could just make them out with the naked eye at a distance of seven and a half miles. There was now an overall covering of nine or twelve inches of snow on the tops, not quite burying the old heather. There were no ptarmigan below Carn Ban, nor could I see any on the smooth white face of the Sgoran which, flawlessly mantled with snow and wreathed in mist, loomed above me in remote and gigantic inaccessibility. All life was centred around the corries where, in traversing across one to a herd of one hundred and thirty-eight stags, I put up four ptarmigan, a grouse, a raven and a big fox, all bushy tail.

February 14 was my thirty-sixth and last expedition to the tops during our first two years at Drumguish, for the earliest heavy snow of the winter began to fall on the unusually late date of the 19th. Although I went up with skis a week later, the snow was frozen too hard for me to make any progress without crampons. Thereafter, deep snow and hard frosts isolated the hills for twenty-seven consecutive days, and my contact with them was restricted to watching through binoculars the companies of stags marching along their dazzling white ridges.

28 Reindeer in the Cairngorms

ON A NOVEMBER day reindeer were browsing among the thick heather and sphagnum north of Loch Morlich: a big grey ox, a young ox with a gleaming yellow eye and a chocolate-coloured coat as glossy as sealskin, three cows and three calves. They were some of the twenty-nine reindeer imported to Glen More in 1952 and subsequent years from Swedish Lapland and Norway. Despite the big ox's impressive spread of antlers these reins looked very small and lightly framed in comparison with red deer: yet the bulls weighed between sixteen and twenty stone, which was markedly heavier than the average contemporary Highland stag. Stags all rangy leg, muscle and sinew: reindeer compact and plodding.

The mists had lifted from the morning gloom to reveal a pale blue sky, the dark silhouettes of long brown hills, and corries lightly dusted with snow. The low sun, streaming through the dark green pines and over the dull red heather and bleached mosses, lit a silky sheen on the greyish coats of the reindeer calves. A piece of Lapland, it had been said, this 300-acre corral of rank old heather, sphagnum-bog and peat-moss studded with old pines, and framed by the stony ridges of the Cairngorms.

Reindeer, in their glossy autumn pelage, were barely recognisable as the same animals during their spring moult and, as it happened, my next encounter with them was on a sunny April afternoon. We were just about to rest at a height of 3,000 feet, after climbing the steep Fiacaill ridge above Loch Morlich, when I became aware that among the confusion of granite blocks all around us were some strange animals, whose sober colours faded them into the rocks. Not deer, surely? One had not seen much of deer in the vicinity of Glen More since it had become a metropolis of skiers, climbers, hostellers and commandos, though Corrie an Lochain and Corrie an t'Sneachda had been traditional rutting corries ten years earlier. No, these were reindeer which, after the disastrous initial policy of fencing them in that fly- and midge-infested corral on the shores of Loch Morlich, were now free to range where they would to the 4,000-foot summits of Cairngorm and Cairn Lochan. During the rut indeed the bulls were liable to stampede the cows far beyond the limits of their herded bounds to the remotest parts of Ben MacDhui; and rather than seek shelter in the glen during the winter storms they often preferred to sit out a blizzard on some exposed hill — as a herd of stags would sometimes do, though the latter usually climbed up to the tops *after* a storm and basked in the alpine sunshine.

It was hot out of the wind, too hot perhaps for the reindeer, which were standing apathetically in a hollow among the rocks, their flanks heaving as they breathed heavily through their extraordinarily large, flared nostrils. The three cows would be feeling the burden of their calves, due in the latter half of May, a month and more before the earliest red deer calves, though both deer rutted in October. All five reins were certainly obligingly indifferent to the sudden appearance of two men, being accustomed to regular rounds by their herdman and no doubt satiated with the human contact of the thousands of skiers that invaded their sanctuary during the winter and spring. The older of the two oxen did not evince the slightest sign of uneasiness when we sat down within twelve feet of him, while I changed a film in my camera; and he remained comfortably couched on his crisp mat of fringe-moss and crowberry, laced here and there by a trailing green runner of alpine club-moss, in a pocket among the rocks. If only the red deer had been as phlegmatic what a series of photographs I might have collected.

Reindeer, like red deer, were out of condition in the spring, and these five ragged, nondescript reins resembled heavy-headed goats, or takin, rather than deer. However, the older white-faced ox was in process of growing a luxuriant silver-grey summer coat and the well-developed knobs of his antlers were covered with dark grey velvet. Reindeer exhibited a motley range of colouring, and the oldest of the three cows was a patchy mole-brown, with white anklets ringing her large splayed hooves.

Having obtained an interesting series of photographs of these co-operative beasts, all that was required to perfect this unexpected encounter was a portrait of the impassive older ox; but though he was prepared to accept me at a distance of twelve feet, nine feet was a little too close and, heaving himself up reluctantly, he wandered slowly away over the scree after the cows, which were nibbling here and there as they went. Reindeer are delicate feeders and, rather than nibbling, they appeared merely to brush the herbage with their lips, taking only the tips of the prostrate lichens, dwarf blaeberry and creeping willows – and no doubt the green tips of heather and sedge on the moors.

I followed them over the ridge a short way for one more shot, and after that there was only the dull tinkle of the cows' bells as the finale to my eighty-ninth and last day on the high tops of the Grampians.

Index